BUILDER'S GUIDE TO stucco LATH & PLASTER

Max Schwartz

with

*Walter F. Pruter
Contributing Editor*

Builder's Book, Inc.
BOOKSTORE ■ PUBLISHER
8001 Canoga Avenue / Canoga Park, CA 91304
1-800-273-7375 / www.buildersbook.com

BUILDER'S GUIDE TO STUCCO, LATH & PLASTER

© 2007 Builder's Book, Inc. All rights reserved.

No part of this book may be reproduced or utilized in any form or by any means, electronic or mechanical, including photocopying, recording or by any information storage and retrieval systems, without special permission in writing from the publisher. The information contained in this book is subject to change without notice.

ACKNOWLEDGMENTS / CREDITS

The author and publisher wish to acknowledge Mr. Walter F. Pruter for his generous contributions to this volume. His review of the book, thoughtful suggestions and numerous contributions helped us create a much more useful reference.

We also wish to extend our appreciation to the Northwest Ceiling & Wall Bureau for their kind permission to reproduce the stucco and stucco-related details that appear in Appendix E. The details we reprint here are but a fraction of the details provided in their *Stucco Resource Guide* (Third Edition, 2002). This is an excellent resource which we highly recommend to anyone wishing to take their knowledge of stucco to the next level.

We would also like to thank the many other sources which inspired many of the illustrations that we created for this book. (See individual figures for credit information.)

NOTICE TO THE READER / DISCLAIMER

This book is designed to provide general information about the planning, selection, use and application of stucco, lath and plaster. The publisher has made every effort to provide complete and accurate information, but does not guarantee the accuracy or completeness of any information published herein. The publisher shall have neither liability nor responsibility to any person or entity for any errors, omissions, or damages of any kind (direct, indirect or consequential) arising out of use of or reliance on this information. This book is published with the understanding the publisher is not attempting to render professional services. If such services are required, the assistance of an appropriate professional should be sought.

ISBN-10: 1-889892-72-6
ISBN-13: 978-1-889892-72-6

For future updates, errata, amendments and other
changes, contact Builder's Book, Inc., 1-800-273-7375.

Contents

1. Stucco Basics 1
    What is Stucco? 1
        The "Foundation" 2
        Base/Subsurface 2
        Plaster Coats 4
        Stucco Application Checklist 6

2. Base/Substrate Preparation and Flashing 9
        Importance of the Base 9
        Are You Ready? 10
    Masonry/Concrete Construction 11
    Wood/Metal Frame Construction 18
    Flashing 22
    "Accessories" Are Not Optional 28
        Galvanized vs. Zinc 31
        Metals vs. Plastics 31
        Seals and Caulking 31
        Control Joints 32
        Beads 36
        Grounds / Base Screeds / Weep Screeds 36
    Integration of Accessories, Lath and Barrier 37
    The Weather-Resistant Barrier 43
        Felt, Kraft Paper or House Wrap? 43
        Installing the Barrier 46
    Lath 49
        Metal Lath to Wood Framing 49
        Metal Lath to Metal Framing (Using Screws) 50
        Metal Lath to Concrete and Concrete Masonry 50
    Exterior Insulation and Finish Systems (EIFS) 55
    One-Coat Stucco Systems 60

3. Plaster Mixes 63
        Different Coats = Different Mixes 63
        Portland Cement 63
        Sand 64
        Water 64
        Precautions While Working with Cement 65
        Mixing Scratch & Brown Coats 66
        Mixing the Finish Coat 68

4. Applying Plaster 71
        How Many Coats? 71
        Is the Weather Right? 72

    Are You Ready?... 73
    Hand or Machine Application?... 74
      Machine Placement and Set-Up.. 75
    Application Patterns.. 80
    Application Techniques... 81
      The First (Scratch) Coat... 81
      The Second (Brown) Coat... 83
      The Final (Finish/Texture) Coat... 86
    How Thick?... 86

5. Plaster Finishes 89
    The Final (Finish/Texture/Color) Coat... 89
    Preparing the Surface 90
    Texture Finish Choices.. 90
      Achieving Finish Textures 96
      Built-In Color 98
      To Paint or Not to Paint?... 98

6. Decorative Stucco 101
      Shapes attached to stucco (plant-Ons or Implants) 102

7. Properties of Plaster 105
    Structural Strength... 105
      Plaster in the Northridge Earthquake... 109
    Thermal Properties... 112
      Insulating Qualities 112
      Fire-Resistive Properties 113
    Water-Resistance.. 121
    Acoustic Properties.. 122
      The Acoustics of Typical Walls.. 123

8. Cost Estimating 125

9. Stucco Maintenance and Repair 135
    Identifying Damage 135
    Paper and Lath Repair.. 137
    Cleaning Plaster.. 141
    Refinishing or Changing the Color.. 143

10. Glossary 145

11. Associations and Major Manufacturers 157

12. References / Bibliography 165

APPENDIXES

A. History of Stucco 169

B. Cement Manufacture 179

C. Governing Codes and Specifications 185

D. Scaffolding and Safety 231

E. Stucco and Related Details 247

INDEX 279

CHAPTER 1

Stucco Basics

WHAT IS STUCCO?

Stucco–also called Portland cement plaster–is an attractive and durable cement-based coating that can be applied over masonry, properly prepared metal or wood framing and/or sheathing, or special insulation board surfaces.

Stucco can be applied to both interior and exterior wall surfaces; this book, however, is mainly concerned with the exterior use of stucco. Keep in mind that the same techniques and materials can be used in interior spaces where moisture resistance, a relatively rough/rustic look and a low-maintenance finish is desired.

Stucco makes an excellent exterior surface. As mentioned before, it is durable and resists weather. It provides some level of soundproofing, and it can withstand moderate earthquakes with minor cracking. And stucco provides good wind protection. For these reasons and more–even in severe climates–you'll find many older stucco buildings still in great condition.

When it comes to aesthetics, depending on the finish texture you use, stucco can create great interest with highlights and shadows both in daylight and at night with property lighting. And when it comes to color, stucco offers the choice of long-lasting, "built-in" color, or it can be painted.

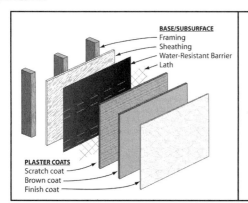

Figure 1-1
Typical 3-Coat Stucco System

A stucco finish is really two or three thin coats of a mortar–also called "plaster"–that is generally one part Portland cement to three parts sand, with a small amount of lime and water added.

> **NOTE**
>
> *In the strictest, technical terms, "stucco" refers to the final ("finish") coat of exterior plaster in which the final color and texture are obtained.*
>
> *In common usage, however, the term "stucco" is used to refer not just to the finish coat, but any or all of the multiple layers of Portland cement plaster, and also to any exterior finish "system" that incorporates a type of lath covered in one or more coats of cement plaster.*
>
> *In this book, we'll use the word "stucco" more inclusively: We will use "stucco" to refer to any or all of the layers in a stucco installation, as well as the entire system itself.*

Although stucco at first seems like a very simple technique, inadequate preparation, incorrect or inappropriate mixtures, or improper application can result in major problems causing the plaster to bulge, separate, crack, or worse: allowing the entry of water, causing the wooden subsurfaces and frames to rot or develop mold or mildew.

In the next few pages, let's briefly review the steps to creating an exterior stucco finish. In the chapters that follow, we'll cover each of these topics in greater detail.

THE "FOUNDATION"

Just as with any other aspect of construction, your finished product is only as good as your base or foundation. In the case of stucco, the base/subsurface is the foundation.

The proper application of stucco requires some care and attention to preparation of the underlying structure and base/subsurface. The entire process involves several distinct steps, none of which should be overlooked.

- The building frame/structure should be stable and solid; any shifting, settling or twisting of the finished stucco surfaces can result in cracks or even complete failure.

Stucco-finished walls (like most shear walls) can add limited strength to a building's structure, but they most certainly cannot make up for shoddy construction.

BASE/SUBSURFACE

Different substrates dictate different approaches to creating a suitable base or subsurface for the coats of cement plaster that are to follow.

METAL AND WOOD FRAME BUILDINGS

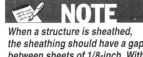
When a structure is sheathed, the sheathing should have a gap between sheets of 1/8-inch. With this gap, should the sheathing become wet, it can swell/expand without causing severe cracking of the stucco.

- In most one-story buildings, the subsurface materials (water resistant barrier and lath) can be attached to the frame structure without plywood or OSB sheathing, unless seismic or wind resistance values are desired. In any event, the inclusion of any rigid sheathing typically improves the quality of the plaster installation.

- In multi-story buildings, sheathing is usually required on the first floor exterior wall surface before the subsurface materials are attached. (Attach the lath to the frame–not the sheathing.) Better quality construction will use the sheathing regardless.

CONCRETE/MASONRY BUILDINGS

- In the case of masonry or concrete walls, you can apply stucco directly to the base (without a weather-resistant barrier or lath), using a bonding agent.

- In some cases (e.g., when the concrete/masonry surface is too rough or when it has been treated with sealants or other finishes that prevent a good bond), you will need to attach self-furred metal lath to the concrete/masonry base.

EXTERIOR INSULATION AND FINISH SYSTEMS (EIFS)

- As we will see in Chapter 2, each EIFS system is different. Generally, we can say that successful EIFS systems require close adherence to the manufacturer's instructions.

ONE-COAT BASE COAT SYSTEM

- As we will also see in Chapter 2, One-Coat Stucco Systems are a recently developed, low cost, imitation stucco system that provide greater insulation and speedier installation. As systems, they must be evaluated by code authorities and building officials before being accepted in lieu of conventional plaster or EIFS.

WEATHER-RESISTANT BARRIER & LATH

- The weather-resistant barrier can be provided by a housewrap, building paper or building felt, although felt is seldom used for several reasons.
- The number of layers to the weather-resistant barrier (one or two) and the types of materials you may use will depend to a great extent to local building codes.
- The weather-resistant barrier must be installed in a shingle-like fashion so that any moisture that enters behind the stucco finish will be guided down and safely away from the interior of the building.
- Lath "accessories"–casing beads, weep screeds, corners, etc.–should be installed with caulking for optimum moisture resistance.
- The weather-resistant barrier, lath accessories and the building's flashing should be viewed and installed as an integrated system that works together to protect the structure and its interior from water intrusion.
- The lath, fastened securely to the frame of the underlying structure, provides the link between the structure and the Portland cement plaster coat(s).
- Properly installed lath laterally reinforces the plaster basecoat and can minimize cracking.
- The lath should be consistently "furred," meaning that the lath should be held away from the plane of the weather-resistant barrier to allow the encasement of the lath within the cement plaster. To achieve this, furring nails can be used to attach the lath, or self-furring lath can be used.
- Individual sheets of lath should be adequately overlapped and wired together to create a continuous layer of reinforcement.
- The accessories should be wire tied to lath wherever they meet.

⚠ CAUTION

Take care in mixing the plaster because the proportions of ingredients are very important for a satisfactory job. You can mix your own mortar or buy it premixed. For smaller jobs or if you are inexperienced, however, it is advisable to buy premixed stucco.

PLASTER COATS

Once the base/subsurface is prepared, it's on to mixing, applying and curing each of the plaster coats. Stucco can be applied by hand or by machine.

Portland cement plaster is mixed on the job by combining plastering sand, Portland cement, a plasticizing agent, and enough water to produce a consistency suitable for the application method you'll be using. Consideration must be given to selecting a plaster mix that can accommodate the maximum allowance quantity of sand.

Timing is everything when it comes to applying each stucco coat, but relative humidity can be just as important. Low humidity and hot, dry winds increase the likelihood of too-rapid evaporation of water from the fresh plaster.

- If at all possible, choose an overcast day to apply any of the stucco coats to walls that have southern exposure.
- The ideal temperature for installing stucco is between 50 and 80 degrees Fahrenheit, regardless of exposure.
- Excessive heat can accelerate the drying process and cause shrinking and cracking.
- Too-cold temperatures, on the other hand, can make the stucco mix too stiff for proper troweling.
- Just prior to applying the first coat over a concrete/masonry base, saturate the wall thoroughly and allow the surface to dry (this is called "Saturated/Surface Dry" or "SSD") to prevent the base from drawing out the cement plaster's moisture too quickly.
- If you are applying a liquid bonding agent, read and follow the manufacturer's recommendations.

Depending on the base/substrate and other factors, a stucco finish may consist of one, two or three coats. Each coat serves a different purpose. In a three-coat system, for example, the three coats in the order of their application are:

- Scratch Coat (or Base Coat): The first coat is applied directly to lath over the substrate (in frame construction, the weather-resistant barrier on sheathing or over open stud framing). A properly applied scratch coat places the plaster behind and around the lath/mesh. Once the plaster has initially set but before it has dried, it is scored horizontally with a scratching tool to produce lines approximately 1/8" deep, to provide "keys" for the next coat and to retain moisture. This first coat must be kept continuously damp for at least 48 hours unless the second (brown) coat is applied as soon as the scratch coat is rigid enough to accept it.
- Brown Coat: The second coat is applied over a damp scratch coat to level the surface and bring the total thickness to grounds, leaving flat and true to plane with no deviation more than ¼" when measured under a 5'-0" straight edge.

- Finish/Texture Coat: The texture and color is established with the final coat. It may be a factory-blended or site-mixed blend of white or gray cement, blended with lime, oxide pigments and aggregate to produce a 1/8" finish coat that can be troweled, floated, brushed, sprayed–or a combination of these.
- Following each coat, a curing process takes place.
- Mist the stucco coat lightly one or more times over the cure period, to allow for a slow, smooth drying process. Alternatively, you can loosely wrap the building with plastic sheeting to retain moisture. In summer heat, you may need to do both.
- Lightly mist the stucco scratch coat just prior to applying the brown coat, and mist the brown coat just prior to applying the finish coat.

As an alternative, the brown coat may be applied as soon as the scratch coat is rigid enough to receive the brown coat. When using this method, calcium aluminate cement (up to 15 percent of the weight of the Portland cement) is added to the mix (see 2508A.6 "Alternate Method of Application" on page 187.)

IMPORTANT

This book presents general guidelines and suggestions for a successful stucco finish, based on common knowledge at the time of this book's publication. No book can address all of the unique specifics of your project or the combination of materials that you choose to use.

- Information in up-to-date manufacturers' documentation takes precedence over anything you read here. That's why it's important to obtain, read, understand and follow all available documentation for each product that your stucco project will use.
- Local codes and ordinances may prohibit some of the techniques or materials discussed in this book. Before you commit time or resources to a particular approach, you must check to make sure that your project design and your chosen materials and techniques will satisfy local requirements.

STUCCO APPLICATION CHECKLIST (RESIDENTIAL)

Project Name	
Address	
City/State/Zip	
Owner	
General Contractor	
Stucco Contractor	

	YES	NO	N/A	COMMENTS/NOTES
PRE-START				
Pre-installation construction conference				
Any details need to be discussed with other trades before proceeding? (Describe on separate sheet.)				
Any issues not covered in this checklist? (Attach sketches or list as applicable.)				
SUBSTRATE				
Sheathing Type:				
Plywood/OSB/Bildrite has 1/8-inch gap between panels				
Sheathing is correctly fastened				
BUILDING PAPER INSTALLATION				
Housewrap used? (Provide manufacturer and specifications.)				
2 layers of Type I Grade building paper used?				
Paper wraps internal and external corners				
Paper lapped shingle style for positive drainage				
Vertical joints lapped a minimum 6 inches				
Horizontal joints lapped a minimum of 2 inches				
All breaches created by penetrations sealed				
Building paper is continuous to underside of rafter or truss top chord				
ROOF LINE				
Chimney cap flashing extends over stucco on stucco chimney(s)				
Interface of chimney cap flashing and stucco is sealed				
Cricket/saddle is installed and properly flashed on high side of chimney				
Wall/roof intersections appropriately flashed				
Kick-out flashings installed				
Stucco is stopped and spaced above roof with an accessory				
Gutters to be installed				

	YES	NO	N/A	COMMENTS/NOTES
WINDOWS / DOORS				
Manufacturer(s):				
Window mfr's installation instructions followed				
Rough openings prepared and panned prior to installation of windows and doors				
Drip cap / head flashing installed				
Stucco stopped with a casing bead around window				
3/8" gap between jamb/sill and stucco				
Closed-cell backer rod and sealant installed in gap around sills and jambs				
Type and brand of sealant(s) used:				
DECK				
Deck attachment properly flashed				
End dam flashing fabricated and sealed where deck stops within stucco wall				
Handrails attached over stucco				
ACCESSORIES AND OTHER ISSUES				
Stucco is terminated with a weep screed 4-inches above grade or 2-inches above pavement				
Quoins, bands and other decorative elements installed after the stucco brown coat				
Pipe sleeves, fixtures, outlets, downspout attachments vents and other attachments sealed				
Flashing installed at transition of stucco and other material				
Vertical transition from stucco to other cladding material sealed				
Projecting wood trim flashed				
Capillary break and weep provided where stucco columns rest on concrete pad				
LATHING				
Proper metal lath used				
Lath attachment fasteners observe proper spacing				
Lath lapped adequately at sides and at ends				
Fasteners penetrate framing 3/4" minimum				
Lath returned around corners or Corneraid/Cornerite used				
Lath applied with long dimension at a right angle to the framing				
Ends of lath adequately staggered				
Self furring dimples project inward				

	YES	NO	N/A	COMMENTS/NOTES
Cups or teeth of expanded metal lath project upward				
Lathing inspected				Inspector: Date:
STUCCO				
Acrylic admixture added to scratch and brown mix				
Minimum temperature maintained during stucco installation and 48 hours after (40° F)				
Total of scratch + brown coats = ¾-inch thick				
Brown coat cured prior to installation of finish				Number of days in cure:
Acrylic finish used				
Manufacturer brand & type:				
Manufactured one-coat stucco used in this installation?				If so, was manufacturer's installation card filed with building official?
NOTES / COMMENTS				

Adapted from "Residential Stucco Installation List," Minnesota Lath & Plaster Bureau

CHAPTER 2

Base/Substrate Preparation and Flashing

IMPORTANCE OF THE BASE

Building structures gain much of their strength from the combination of bracing (the "frame") and shear walls. Shear strength is vitally important when wind and earthquake are potential problems, but even in those regions where wind and earthquake risks are minimal, providing wall surfaces with adequate shear strength can mean the difference between a strong building and a collapse.

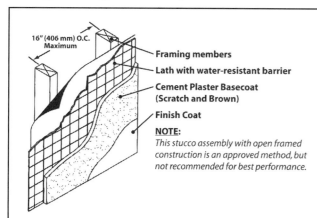

Figure 2-1
Open Stud Construction

Note: In California, in multi-story buildings, wood sheathing is required on first-floor walls, as shown in Fig. 7-2.

NOTE
When a structure is sheathed, the sheathing should have a gap between sheets of 1/8-inch. With this gap, should the sheathing become wet, it can swell/expand without causing severe cracking of the stucco.

In Chapter 7, you'll find a more detailed discussion regarding the strength that is provided against lateral (horizontal) forces by sheathing a building with plywood or OSB panels, versus relying on lath and plaster alone to form the "sheathing." The bottom line is this: applying lath and plaster alone cannot match the strength provided by sheathing a building in plywood/OSB alone. Lath and plaster alone falls even farther short of the combined strength provided by lath and plaster over a plywood/OSB substrate.

In California, following the 1994 Northridge Earthquake, standards were improved so that it is now a requirement that all multi-story buildings provide at least plywood/OSB sheathing on their first floors. For economy's sake, many builder's still forego the wood sheathing on single-floor residences, but better quality builders use plywood/OSB on all exterior wall surfaces.

ARE YOU READY?

- If you will be using scaffolding, it should be in place and secured according to all safety regulations. See Appendix D on Scaffolding.

MASONRY/CONCRETE CONSTRUCTION

A concrete or masonry base allows you fewer steps when it comes to stucco application. First, unlike frame construction, there's no need for a weather-resistant barrier. And the use of lath is optional, as you'll see, depending on the quality of the underlying surface.

Masonry or concrete substrates usually require at least a minimum of preparation to ensure that the plaster establishes a strong bond. Some surfaces will require sandblasting, or they may require mechanically abrasion or other work to remove contaminants and roughen the surface. This will enhance the bonding capability of the new plaster.

Although stucco should bond well to a properly prepared substrate, bonding agents can also be used. All bonding agents should conform to ASTM C 932.

Note: There are interior-grade (less expensive) and exterior-grade (more expensive) bonding agents. Despite being clearly inappropriate, many stucco contractors try to get by with the less expensive interior product. The result, unfortunately, is frequent and sometimes rapid failure in the stucco bond. Use or specify exterior-grade bonding agents to ensure the best results possible.

For example, you can apply stucco directly to relatively smooth concrete, by pre-treating the concrete surface with a good bonding agent. Any paints or sealants that may have been used on existing concrete or masonry could affect the ability of the cement plaster (and even bonding agents) to adhere properly to the base.

Because concrete and masonry are porous, chemically stripping the surface may not extract all remnants of these products, which can be absorbed deeply into the masonry/concrete base. In those cases, metal lath securely attached to the substrate may be required to create a proper surface for stucco.

- If the masonry or concrete is dry, saturate the wall thoroughly and allow the surface to dry (this is called "Saturated/Surface Dry" or "SSD"). This will prevent the base from absorbing an excessive amount of water from the plaster. (If you are using a bonding agent, you likely will not pre-wet the masonry or concrete; read and follow the bonding agent instructions.)
- The same rules that apply to a masonry or concrete substrate also apply to the edges of existing plaster to which patch material will bond. The edges of existing plaster in a patch area should be roughened and properly dampened to SSD.
- In stucco that is applied over concrete or masonry, control joints are only required to be installed where they are installed in the underlying base.

DEFINITION
A stucco finish coat applied directly to masonry or concrete is also called a "skim coat."

HOW-TO'S
Because stucco is a like-kind material to concrete or concrete block substrates, expansion joints in stucco in direct application over these substrates are dictated by whether a joint occurs in the substrate itself.

CAUTION
Not all bonding agents can be applied over smooth concrete; read and follow instructions for the bonding agent you have selected–and remember to use a bonding agent designed for exterior use.

> **DEFINITION**
> A stucco finish coat applied directly to masonry or concrete is also called a "skim coat."

STUCCO OVER SMOOTH CONCRETE

If the underlying concrete is relatively flat, a single, finish coat of stucco applied at approximately 1/8" thick can be used following the bonding agent.

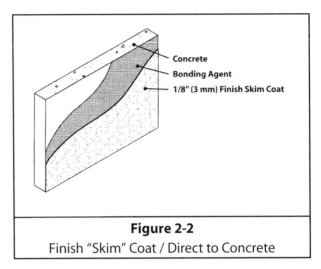

Figure 2-2
Finish "Skim" Coat / Direct to Concrete

- Apply an exterior-grade bonding agent to the concrete surface (optional).
- If no bonding agent is used, saturate the wall thoroughly, then allow the surface to dry (Saturated/Surface Dry).
- Apply a finish/skim coat, either by hand or by machine (pump).
- Texture the finish coat.

If the underlying concrete surface has a gloss finish, it can be roughened up by mechanical means (grinders, etc.). Or, a liquid bonder, bonding coat or dash bond coat may be applied to the surface to allow direct application to proceed.

- A dash or bond coat uses a slurry-consistency plaster mix high in cement content.
- It can be machine-applied, or applied by hand with brushes, trowels or paddles.
- You quite literally "dash" or splatter the slurry onto the surface, so that most, but not all (approximately 60-70%) of the surface is covered.
- Do not smooth the dash coat. Allow it to dry rough. Part of its effectiveness is that it provides a rough surface for the cement plaster coats to follow.
- Following the dash coat, you then apply the finish/skim coat.

STUCCO OVER MODERATELY ROUGH CONCRETE

If the underlying concrete is not smooth (has gaps or chinks, for example), a two-coat stucco finish can be used to mask the underlying cosmetic problems. Alternatively, patching the concrete surface may make the "Stucco Over Smooth Concrete" method possible.

If the surface of the concrete is overly smooth, you may want to consider either mechanically roughening the surface, using a dash or bond coat (as discussed above for Stucco Over Smooth Concrete), or applying an approved liquid bonding agent.

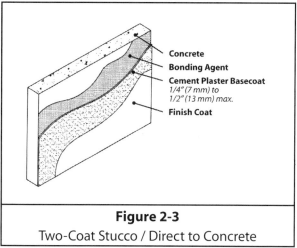

Figure 2-3
Two-Coat Stucco / Direct to Concrete

- An exterior-grade bonding agent is used prior to or along with the first coat, which is then treated like a brown coat. The first coat should be at least ¼" (7 mm) thick and no thicker than ½" (13 mm).
- Dampen the first coat before the finish coat is applied.
- Apply a finish/skim coat, either by hand or by machine (pump).
- Texture the finish coat.

STUCCO OVER PROBLEM CONCRETE

If the underlying concrete presents real problems (stubborn paints or sealants, significant surface irregularities, etc.), you may need to use self-furred metal lath to provide a consistent and sufficient bonding key. This can then be followed with a three-coat stucco application.

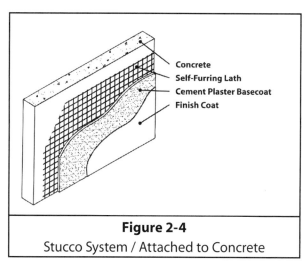

Figure 2-4
Stucco System / Attached to Concrete

- Attach the lath 6″ O.C. in vertical directions, and into supports on horizontal directions.
- Use power-actuated fasteners designed for attaching metal lath to masonry and concrete.
- The first coat will be treated like a scratch coat (three-coat system).
- The first (scratch) coat should be at least 3/8″ (11 mm) thick and no thicker than ½″ (13 mm).
- Allow the first coat to moist cure (don't forget to "scratch" it if you are applying a total of three coats).
- Dampen each coat lightly before the next coat is applied.
- Apply a finish/skim coat, either by hand or by machine (pump).
- Texture the finish coat.

STUCCO OVER CONCRETE AND WOOD BASES

If you are "bridging" across dissimilar substrates (for example, transitioning from a concrete or masonry base to a framed/sheathed base), you can attach metal lath directly to the concrete/masonry surface, and attach a water barrier and lath over the framed/sheathed section. The water-resistant barrier (paper) should overlap the concrete or masonry about 6 inches, under the metal lath.

- Use a control joint along the boundary of the two bases to allow for expansion and contraction and minimize cracks.

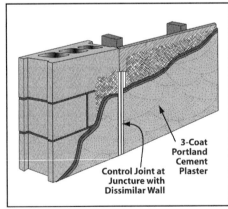

Figure 2-5
A control joint at the boundary between dissimilar base materials. It is recommended that the water-resistant barrier overlaps (extends over) the concrete/masonry by 6 inches, under the metal lath.

Source: The Quikrete Companies, 2005.

- Attach the lath every 6″ O.C. vertically and not over 24″ O.C. horizontally on masonry and into supports on frame construction.
- Attach the self-furred type lath at furring deformations and flat lath with furring nails not over 6″ O.C. Fasteners must be long enough to penetrate supports at least ¾″. Do not attach to sheathing alone.
- On the concrete/masonry portions, use approved power actuated fasteners. (The head is larger than standard masonry nails, ensuring that the lath doesn't pull through.)

- The first (scratch) coat should be at least 5/8" (15 mm).
- Allow the first coat to moist cure (don't forget to "scratch" it).
- Dampen each coat lightly before the next coat is applied.
- Apply a finish/skim coat, either by hand or by machine (spray gun).
- Texture the finish coat.

STUCCO OVER MASONRY

In the case of masonry, the choices are similar to those we've reviewed for solid concrete surfaces. However, because the mortar joints between the masonry units may not be flush with the masonry surface, the surface of a masonry wall often is not amenable to a single, finish stucco coat only (as is possible with a smooth concrete wall). Clearly, if the masonry wall is new, the masons should be instructed to provide flush mortar joints ("struck flush").

On a *new*, struck-flush concrete masonry wall that has adequate absorption, you can usually get by without a bonding agent. Allow the wall to cure thoroughly before beginning the stucco project, however. For a surer result, however, a bonding agent is recommended. A dash or bond coat is not necessary on new CMU walls.

- Apply an exterior-grade bonding agent to the masonry surface (optional).
- Only if the bonding agent calls for it, saturate the wall thoroughly, then allow the surface to dry (Saturated/Surface Dry).
- Apply a first coat (which is treated as a brown coat–not scratched).
- Apply a finish/skim coat, either by hand or by machine (pump).
- Texture the finish coat.

On *existing* masonry surfaces, it's important to consider the type of masonry units that have been used–whether they are concrete masonry units (CMUs) (sometimes referred to as "cinder blocks"), unglazed clay tile, nonabsorptive hard burnt brick–and the type and condition of the mortar between the blocks/bricks.

Where the masonry surface has been treated with water sealers or paints, sandblasting and possibly chemical preparation may be necessary. If the surface is still non-absorptive or uneven, or otherwise unlikely to provide a proper bonding surface, you may be better off attaching metal lath to the masonry base.

- For this purpose, use power actuated fasteners, which are designed specifically for attaching metal lath to masonry surfaces. The head is larger than standard masonry nails, ensuring that the lath doesn't pull through.

If some of the mortar joints are brittle (i.e, they easily crumble), those specific joints should be raked out with a chisel, and repaired (re-pointed)

> **⚠ CAUTION**
> Don't rake out masonry mortar joints simply to provide a "key."
>
> Doing so will create plaster of varying thicknesses and "ghosting" of the joints, which can (at worst) cause cracking, and (at best) an uneven surface.

if necessary. Do not rake mortar joints in masonry simply "to provide a key." Using a dash or bond coat (discussed below) is preferable.

Figure 2-6
Removing Loose or Brittle Mortar

If the underlying masonry materials have a gloss finish, applying a dash or bond coat or liquid bonder to the surface may allow a direct-to-surface application to proceed.

- A dash or bond coat or uses a slurry-consistency plaster mix high in cement content.
- It can be machine-applied, or applied by hand with brushes, trowels or paddles.
- You quite literally "dash" or splatter the slurry onto the surface, so that most, but not all (approximately 60-70%) of the surface is covered.
- Do not smooth the dash coat. Part of its effectiveness is that it provides a rough surface for the cement plaster coats to follow.

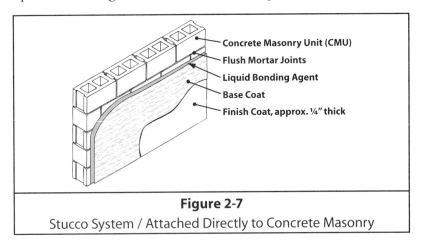

Figure 2-7
Stucco System / Attached Directly to Concrete Masonry

Source: Northwest Wall & Ceiling Bureau, Stucco Resource Guide (Third Edition)

- Depending on the number of stucco coats you'll be applying, the first full coat will be treated like a scratch coat (three-coat system), or like a brown coat (two-coat system).
- The first coat should be at least ¼" (7 mm) thick and no thicker than ½" (13 mm). If you are applying only two coats, the first coat should be on the thick end of this range (3/8" to ½").

- Properly moist cure each basecoat, and don't forget to "scratch" the first coat if you are applying a total of three coats.
- Dampen each coat lightly before the next coat is applied.
- Apply a finish/skim coat, either by hand or by machine (spray gun).
- Texture the finish coat.

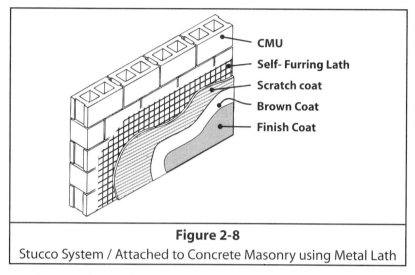

Figure 2-8
Stucco System / Attached to Concrete Masonry using Metal Lath

Source: Northwest Wall & Ceiling Bureau, Stucco Resource Guide (Third Edition)

In some cases where you want or need to provide insulation, you can do so by using steel z-channel furring. The z-channel allows insulation boards to be nearly continuous along the wall surface (with little or no gap between panels.

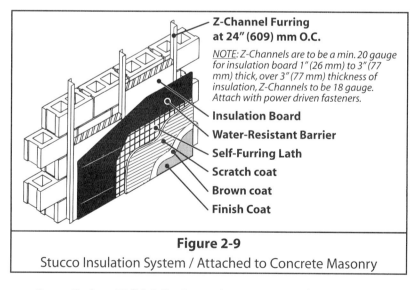

Figure 2-9
Stucco Insulation System / Attached to Concrete Masonry

Source: Northwest Wall & Ceiling Bureau, Stucco Resource Guide (Third Edition)

- Attach the z-channel to the masonry with hardened masonry nails, then insert the insulation panels.
- The sheathing is then attached to the z-channel.

HOW-TO'S

When the structure is sheathed, the sheathing should have a gap between sheets of 1/8-inch. With this gap, should the sheathing become wet, it can swell/expand without causing severe cracking of the stucco.

The building should be carrying 90% of its dead load prior to the installation of the stucco.

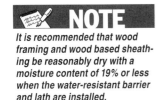
It is recommended that wood framing and wood based sheathing be reasonably dry with a moisture content of 19% or less when the water-resistant barrier and lath are installed.

After that, the stucco installation follows that used for a wood framed structure.

- Attach the lath to the z-channel.

WOOD/METAL FRAME CONSTRUCTION

The frame building can be sheathed or not. If sheathing is used, it can be gypsum, plywood or OSB. However, as we mentioned earlier in this chapter, California now requires that all multi-story buildings provide at least plywood/OSB sheathing on their first floors. That provides a certain level of structural strength that builders in all states should strive for, regardless of local codes.

For economy's sake, many builder's still forego the wood sheathing on single-floor residences (even in California), or on the upper floors of a multi-story building. Better quality builders use plywood, OSB or gypsum sheathing on all vertical wall surfaces, on all stories, because sheathing assures flatter walls and more uniform plaster thickness.

Whenever wood-based sheathing (plywood or OSB) are used, 1/8" gaps should be provided between panels, to allow for expansion.

OPEN FRAME CONSTRUCTION

In those instances where you do not use sheathing over the wood or metal framing, some types of lath require "line wires" under the weather resistant barrier and lath. The wire supports the weather-resistant layer (either building felt or paper), keeping the paper relatively flat. This results in an even thickness of cement plaster between the furred lath and paper backing.

- Use minim 18-gauge, soft annealed, galvanized steel wire for line wires.
- Attach a horizontal line wire every 6".
- Nail or staple each horizontal wire as taut as possible to every fourth stud (#1 and #4 in the figure below).
- As a final measure to stretch or tighten the wire, raise or lower it and secure it to either of the two intervening studs (#2 or #3 in the figure below).

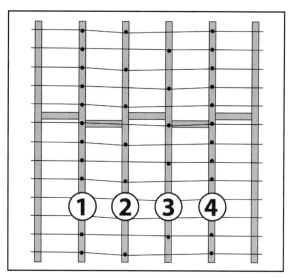

The weather-resistant barrier is adhered to the face of the studs with adhesive or randomly stapled, to temporarily hold it in place. If flat (not self-furred) paper-backed metal lath is used, self-furring nails should be used in wood frame construction. Self-furring screws should be used in metal frame construction. A self-furring fastener comes with a ¼" fiber wad surrounding the shaft.

- Position the wad behind the lath and against the weather-resistant barrier/wall.
- Hook the lath over the fastener head and attach the fastener into the stud.

SHEATHED CONSTRUCTION

Depending on local building codes, you may have a choice of exterior sheathing materials:

- Exterior gypsum sheathing.
- Plywood, OSB, or other wood-based sheathing.
- Expanded (EPS) or Extruded Expanded (XPS) Polystyrene insulation panels.

The weather-resistant barrier and lath are installed over the sheathing. (Most local codes require two layers in the weather-resistant layer, as shown in Figure 2-10, when wood-based sheathing is used.; only one layer of weather-resistant barrier is required when gypsum sheathing is used.)

If flat, paper-backed metal lath is used, you may or may not need to use self-furring fasteners. For example, paper-backed wire fabric lath is nearly always self-furred, so regular fasteners may be used. Paper-backed expanded metal lath is usually self-furred (but not always), so you may or may not need to use self-furred fasteners with that type.

If self-furring fasteners are needed, self-furring nails should be used in wood frame construction, and self-furring screws should be used in metal frame construction. A self-furring fastener comes with a ¼" fiber wad surrounding the shaft.

NOTE

A framing inspection should take place prior to a lath inspection. Appropriateness of flashings, treatments of penetrations and rough openings will be evaluated at that time.

HOW-TO'S

When the structure is sheathed with wood-based panels, the sheathing should have a gap between sheets of 1/8-inch. With this gap, should the sheathing become wet, it can swell/expand without causing severe cracking of the stucco.

NOTE

Plywood and OSB sheathing products used under stucco should meet Exposure 1 or Exterior Grade classifications.

- Position the plug behind the lath and against the weather-resistant barrier/wall.
- Hook the lath over the fastener head and attach the fastener into the stud.

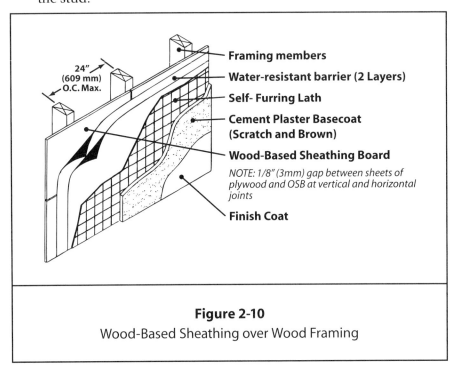

Figure 2-10
Wood-Based Sheathing over Wood Framing

Source: Northwest Wall & Ceiling Bureau, Stucco Resource Guide (Third Edition)

It is recommended that multi-story wood frame construction include expansion joints at floor lines to compensate for wood shrinkage and structural compression. These issues however can be mitigated by engineered structural components that are more stable than dimensional lumber.

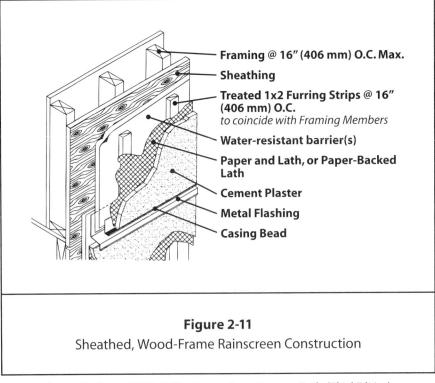

Figure 2-11
Sheathed, Wood-Frame Rainscreen Construction

Source: Northwest Wall & Ceiling Bureau, Stucco Resource Guide (Third Edition)

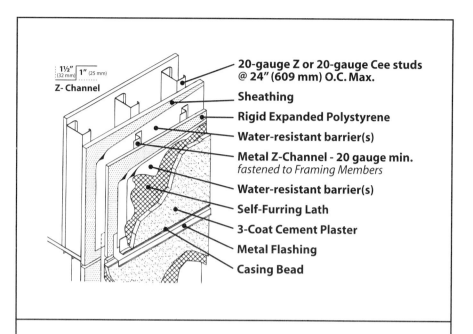

- 1½" (32 mm) / 1" (25 mm) Z-Channel
- 20-gauge Z or 20-gauge Cee studs @ 24" (609 mm) O.C. Max.
- Sheathing
- Rigid Expanded Polystyrene
- Water-resistant barrier(s)
- Metal Z-Channel - 20 gauge min. *fastened to Framing Members*
- Water-resistant barrier(s)
- Self-Furring Lath
- 3-Coat Cement Plaster
- Metal Flashing
- Casing Bead

Figure 2-12
Sheathed, Metal Frame / Rainscreen Construction

NOTE REGARDING RAINSCREEN CONSTRUCTION AND WATER-RESISTANT BARRIERS

Tongue-and-groove Expanded Polystyrene sheathing is considered one layer of weather-resistant barrier, so only one additional layer of barrier can be used to provide redundant protection. Two layers should always be used over wood sheathing, but only one is required over gypsum sheathing.

Rainscreen construction intends that intruding moisture will dam out through the furred space between the back of the EPS and the weather barrier, over the sheathing.

Source: Northwest Wall & Ceiling Bureau, Stucco Resource Guide (Third Edition)

CAUTION

This book is not intended to provide an exhaustive review of how and where to install flashing. There are already a number of good references on this topic, and typically, flashing is not the work of the stucco contractor, but of another (sheet metal contractor, for example). We'll cover the flashing that most concerns the stucco application: flashing around windows and doors.

NOTE

All flashings and water-resistive barrier sheets/layers should be installed for positive drainage.

HOW-TO'S

Wall penetrations such as electrical outlets, plumbing and vents should be completed and properly flashed or integrated to the water-resistive barrier prior to lath and stucco installation.

HOW-TO'S

Joins of flashing sections should be appropriately lapped, sealed and mended together by application of sealants, adhesive-backed membrane or other appropriate means of providing continuity.

FLASHING

This section provides some basic flashing information. Again, local codes or specific project needs may call for specific types of flashing.

There are many alternative methods for designing, fabricating and installing flashing. Most all flashing systems follow a "shingle pattern," however:

- The shingle pattern has all upper layers of weather-resistant protection overlapping onto the lower layers.

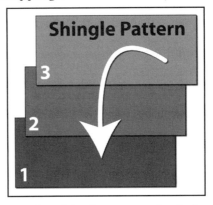

In addition to using the shingle pattern both in flashing and in the weather-resistant barrier layer (which we will talk about in pages that follow), the important thing to remember is that wherever flashing is required, its design should be integrated into the weather-resistant barrier installation and in the lath and lath accessories installation. These are not separate, unrelated systems; they are part of a single system designed to protect a building's structure and interior from damaging water intrusion.

In most cases, you can follow these rules-of-thumb:

- Flashing along top edges of wall surfaces is installed over the weather-resistant barrier layer and under the accessories/lath layer.
- Flashing along the bottom edges of wall surfaces and at the tops of openings (doors or windows) is installed behind the weather-resistant barrier layer.
- Historically, flashing has been accomplished using Grade B waterproof Kraft building paper and sealant. Currently, the most popular method uses "peel-and-stick" rubberized asphalt membranes; they are more water-resistive, effectively seal holes made by fasteners, and eliminate the need for separate sealants.

HEAD FLASHINGS OF WINDOWS

Head flashing is recommended for all windows, although building design (overhangs, soffits, etc.) can make them unnecessary or less-than-critical, or they may be formed with self-adhering, self-sealing flashing membrane. Head flashings are installed behind the weather-resistant barrier.

END DAMS

- End-dams should be installed at head flashing to prevent moisture from entering behind the intersection of the jamb and head.
- The end-dam should be compatible with the head flashing material
- End-dams may be an upturned metal edge or a prefabricated Poly Vinyl Chloride (PVC) accessory, or they may be formed with self-adhering, self-sealing flashing membrane.
- Use caulking when installing the end-dam, to ensure a proper seal.

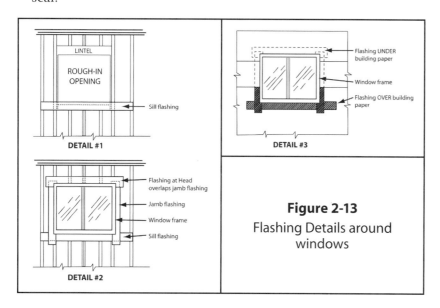

Figure 2-13
Flashing Details around windows

Metal flashing materials should be a minimum 26 gauge galvanized sheet metal or anodized, coil-coated or painted aluminum.

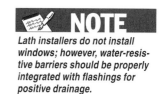

Lath installers do not install windows; however, water-resistive barriers should be properly integrated with flashings for positive drainage.

SILL PANS

Sill pans installed along the bottom of doors and windows should have an up turned back edge and end-dams to prevent moisture from entering the wall cavity.

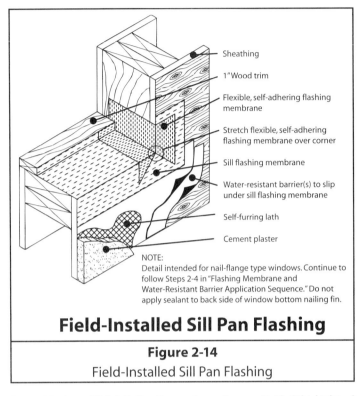

Figure 2-14
Field-Installed Sill Pan Flashing

Source: Northwest Wall & Ceiling Bureau, Stucco Resource Guide (Third Edition)

- End-dams in sill pans may be an upturned metal edge or a prefabricated Poly Vinyl Chloride (PVC) accessory, supplemented with flexible, self-adhering, self-sealing flashing membrane.
- Use caulking when installing the end-dam in a sill pan, to ensure a complete and proper seal.

NOTE

Sections of flashing or trim accessories that intersect or butt-up against each other (at corners, for example) should be lapped, caulked, or have a strip of self-adhering membrane placed over the joints. This prevents moisture from getting to the building structure.

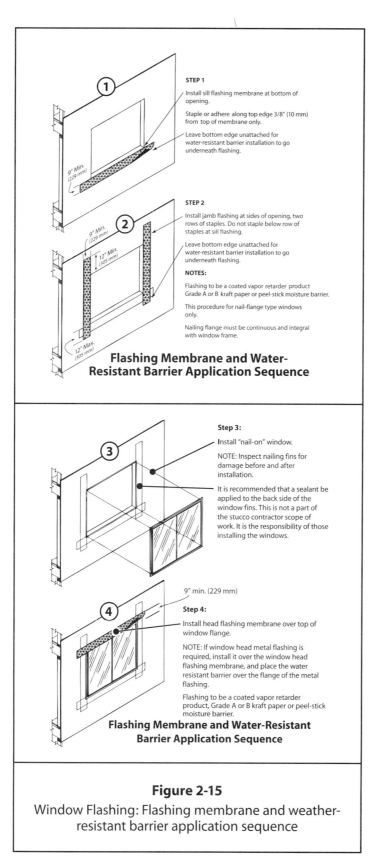

Figure 2-15
Window Flashing: Flashing membrane and weather-resistant barrier application sequence

Source: Northwest Wall & Ceiling Bureau, Stucco Resource Guide (Third Edition)

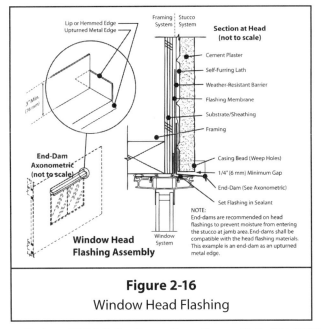

Figure 2-16
Window Head Flashing

Source: Northwest Wall & Ceiling Bureau, Stucco Resource Guide (Third Edition)

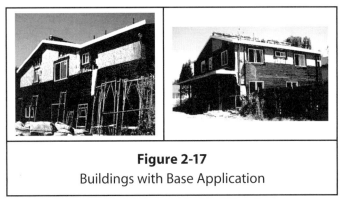

Figure 2-17
Buildings with Base Application

Source: Max Schwartz, 2005.

KICKOUT FLASHING

Design features where different building materials and/or different surface planes meet often are the areas most susceptible to moisture entry and damage; for example, where a lower roof cornice stops in the middle of a stucco wall, or where a dormer wall meets the roofline. In these and similar locations, kickout flashing can make the difference and prevent future water damage.

- Kickout flashing should be fabricated from approved materials, with watertight seams.
- The flashing should be wide enough to handle water run-off from the building's roof pitch.
- Gutters and downspouts should be designed and maintained to handle the runoff from these potential problem areas.
- Kickout flashing pieces are offered in both left- and right-hand versions. Make sure you're using the correct version.

The installation of kickout flashing (like all other flashing) should be coordinated to provide an integrated water protection system.

- On the roof: The kickout flashing is placed on top of the water-resistant barrier on the roof (the roofing felt), and under the roofing material.
- On the wall surface: The kickout flashing is placed behind the weather-resistant barrier on the wall (housewrap and/or approved building paper layers).

As you can see, installing kickout flashing after the installation of the roofing material doesn't make sense, and only creates the possibility for inadequate flashing, damaged roofing materials, and extra labor. What this means in practice is that the kick-out flashing will be installed by the roofing contractor as roofing materials are being laid.

"ACCESSORIES" ARE NOT OPTIONAL

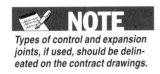
NOTE
Types of control and expansion joints, if used, should be delineated on the contract drawings.

The term "accessories" is used to refer to metal or plastic assemblies that are used to provide proper seals and edges for the cement plaster.

Lath accessories are used in both interior and exterior plaster work. When selecting/purchasing lath accessories, be certain that you have chosen items that are sized and designed for the needs of exterior stucco work. For example, exterior stucco work involves coats of cement plaster that are significantly thicker than the coats of plaster that are used in interior work, and lath accessories come in various sizes and dimensions to accommodate those different needs.

ASTM C 1063 provides specifications and installation guidelines regarding the lath and accessories used with exterior stucco.

- For example, it requires that lath accessories must be fabricated from zinc alloy (99-percent pure zinc), galvanized (zinc-coated) steel complying with ASTM A653, G60, rigid PVC, CPVC or anodized aluminum alloy.

- The exact type of material you select for the lath accessories you use should be determined by the climatic and environmental conditions specific to your project location. For example, extended periods of frost, salt air, industrial pollution, high rainfall or humidity will lead you to choose metal over plastic, plastic over metal, or aluminum over zinc.

Lath accessories are most typically made from and most generally available in zinc or zinc-galvanized steel versions. (Illustrations of the most popular accessories are provided in Table 2-1 below.)

CASING BEADS

Sometimes referred to as "stop beads," "#66 beads" or "J" channels–casing beads are used when it is necessary to terminate the stucco with a clean edge at various locations on the wall plane. These components come in different ground sizes to accommodate stucco thicknesses from ¼-inch to 1½-inch.

Casing beads come in two different basic versions: one has an expanded metal flange, and the other is solid sheet metal with perforated holes. The expanded metal flange is often used because it is easily integrated with the lath and provides an instant mechanical key for the stucco that will follow. The perforated version is more often used when the bead will be attached directly to the substrate with fasteners.

WEEP SCREEDS

These are generally used as a basis or starting point at the mud sill or plate at the bottom of a wall surface that will be receiving the stucco application. As the name implies, the purpose of the weep screed component is to "weep" or pass any incidental moisture that may get behind the stucco to the outside plane of the wall assembly. Contrary to popular belief, however, the perforations at the bottom of a weep screed do little to help evacuate incidental moisture as the holes are typically plugged with stucco. In fact, the holes in weep screeds are more effective at mechanically keying the stucco to the component than providing drainage. Instead, any incidental moisture is more likely to find its way out over the cant of the inverted "V" of the component.

The true function of (and necessity for) weep screeds has generated a great deal of controversy over the years. By most accounts, incidental moisture will dissipate to such an extent at or near its source, that it will never be evident at the point of exit at the weep screed. From most experts' experience, the more important function of a weep screed is its action as a capillary break from the soil and also as a "marker" in identifying where the backfill must stop in the construction.

NOTE
The cement plaster that is applied against the weep screed contracts slightly upon drying, leaving sufficient space for water to escape out over the nose. (It has been found that foundation weep screeds with holes don't perform as well because the holes become plugged with plaster.)

CORNER BEADS

Corner beads are attached to outside wall corners, to true the Portland cement plaster to plumb and straight, as well as to provide strength at these vulnerable locations. Typically corner beads come with various nosings or grounds to screed the plaster to various thicknesses.

In many jurisdictions, corner beads are primarily used for interior plastering applications rather than exterior, because on exterior applications they have a propensity for rusting and cracking along either side of the nosing. Instead of corner beads, the product preferred by experienced plasterers for adding strength to exterior, outside corners are Corneraid or Cornerite. These products are welded wire and are fully embedded in the stucco. If you are using the welded wire product, a certain amount of finesse is required to true a corner with fresh stucco.

CONTROL JOINTS

Control Joints are an artificial means of introducing "controlled stress" at a specific location on a wall plane, in the anticipation that cracking will less likely occur randomly within a stucco panel. Control joints are readily recognized as one piece components with a typical "M" or accordion type shape and expanded metal wings.

Control joints provide for minimal movement capability due to the expansion and contraction of the stucco membrane such as shrinkage in its initial cure. The accordion shape reacts much like a bellows opens and closes in accepting this movement. Several varieties of this component are offered: The "M" shaped profile, sometimes referred to as the "Double V," is perhaps the most common control joint.

In recent years the Double V joint has been largely replaced by the "J Type Joint." The "J Type" includes a lip or ground that provides for a mechanical keying of the stucco at its interfacing edge. This improvement was determined when it was realized that the "M" had a tendency to crack along its sides as the joint opens and closes with stucco movement.

Expansion joints when delineated on the contract drawings, should be located at points where significant building movement is anticipated: for example, at wall penetrations, structural plate lines, junctures of dissimilar substrates, existing construction joints (such as in brick or block), at columns and cantilevers.

It is recommended that multistory wood frame construction include expansion joints at floor lines to compensate for wood shrinkage and structural compression. These issues however can be mitigated by engineered structural components that are more stable than dimensional lumber.

HOW-TO'S

Because stucco is a like-kind material to concrete or concrete block substrates, expansion joints in stucco in direct application over these substrates are dictated by whether a joint occurs in the substrate itself.

EXPANSION JOINTS

While industry standards have not, until recently, addressed the differences between control and expansion joints, it is widely accepted that control joints are single-piece components and expansion joints are multiple-piece components; as a result, expansion joints theoretically offer much more movement capability.

The most popular expansion joint is commonly referred to as the No. 40 Joint. The No. 40 expansion joint has two solid sheet metal sleeves that move independent of each other. Perforations in each sleeve are used to fasten the joint on either side of an isolation joint. In some cases this joint is expanded by the insertion of an additional piece of sheet metal between these sleeves to allow for even greater movement capability, or to provide a larger reveal for aesthetic purposes.

Another recognized method for making an expansion joint is to use two casing beads mounted in a back-to-back or knuckle-to-knuckle orientation. The space between the adjacent joints is generally determined by the anticipated movement of the joint (usually no more than ¾-inch). The space is then filled with a bond breaker and elastomeric sealant.

MISCELLANEOUS ACCESSORIES / SPECIALTY LATH

"Reentrant corners" are areas on a building where there is a natural stress point for stucco to crack. (The most common example of reentrant corners are the corners created by a window opening.) Although it is not mandatory, strip lath (which is usually considered an "accessory") is often employed as reinforcement for these areas. Strip lath is no different than any other expanded metal lath, except that this product is pre-cut into strips approximately 4-6 inches wide.

Similar in use as corner aids or corner bead on outside corners, corner lath is sometimes used on inside corners in lieu of bending lath around the corner. This provides continuity of lath for a better key, and additional strength.

"Reveals" are sometimes used in plastering applications to provide aesthetic rustication (or grooves). A number of manufacturers offer a variety of configurations that are referred to as "Channel Screeds" for banding effects, "F-reveals" for top-of-wall rustication, and other specialty profiles. Often these channel screeds (when designed properly) can also provide the functional performance of a control joint and help control cracking caused by thermal changes and minor building movement.

Manufacturers also offer component profiles for Soffit Vents. Some soffit vents come with built-in drip screeds that isolate the fascia from the soffit. This is helpful in preventing water draining down the fascia from backwashing onto the soffit because of surface tension.

TIE WIRE

Tie wire is the predominant product used to attach metal accessories to lath or lath to cross-furring or gridwork in a suspended ceiling applications.

- The gauge of wire used correlates with the United States Steel Wire Gauge numbering system.

- Wire should be galvanized and annealed low-carbon steel.

Lathers use a variety of tying configurations for specific uses some of which are the saddle tie, butterfly tie, figure eight tie, stub tie and double wrap tie.

GALVANIZED VS. ZINC

Metal accessories are offered in both galvanized steel and solid zinc versions. Since galvanization is simply the coating of steel with zinc, it is reasoned that solid zinc components will better resist corrosion. While this cannot be disputed, you should understand the disadvantages of solid zinc, as well.

For example, zinc is a much more malleable metal than steel, and it can be easily cut by an errant trowel. It is also lighter and more easily bendable, which makes it somewhat more difficult to install. Finally, experience has shown that solid zinc may actually have a greater tendency to expand and contract with thermal changes, potentially resulting in pinched joints and intersections and cracking along edges that may ultimately lead to moisture intrusion issues.

METALS VS. PLASTICS

Galvanized steel and solid zinc components predominate the stucco industry; however, anodized finished aluminum is often used for flashing elements as well as in the reveals described earlier. A word of caution: unless it is protected by some type of coating or gasketing, aluminum is subject to galvanic corrosion in stucco applications.

Many of the same components available in galvanized or zinc are also available in plastic versions. Plastic accessories have their limitations, as they do not perform well in colder climates. Indeed, the predominant reason for their lack of use is the question of whether they can accept the extreme thermal changes. Depending on their quality, they may be susceptible to cracking. bending or splitting..

SEALS AND CAULKING

Before we discuss any of the specific accessories and their uses, we should recognize that joints, intersections and attachment points in lath accessories all create potential sites for water intrusion. You can overcome this potential by following these practices:

- Install each accessory with a strip of building paper into the field of the wall, behind the accessory's entire length. The strip should be 6" to 9" wide.

    On bottom and side edges, the main layer of the building paper will overlap this smaller paper strip, and in a shingle effect, the two will prevent water intrusion to the wall base behind.

    On top edges, the main layer of paper goes behind the smaller strip, with the smaller strip becoming the top-most "shingle."

- Install all accessories in conjunction with flashing or self-adhering membrane.

    Remember, preparation/installation of the stucco base (weather-resistant barrier and lath) and the flashing should be treated as a single, integrated system that protects against water intrusion.

- Use a good quality caulking at joints, intersections, butt ends and along the length of longer accessories prior to their attachment.

    The caulk should adhere to the accessory material as well as to the base.

CONTROL JOINTS

You can minimize cracking in the finished stucco by careful preparation, application and curing of the stucco coats. But even the best stucco can crack, due to factors like these:

- Shrinkage stress
- Building (framing) movements
- Settling foundations
- Construction joints
- Intersecting walls or ceilings, corners, and pilasters
- Restraints from lighting and plumbing fixtures
- Weak sections due to cross-section changes, such as openings

Cement plaster, like concrete and all other cement-based materials, will shrink slightly when it dries. In typical concrete, this change amounts to about 1/16-inch over 10 feet (.4 cm in 3 meters). To relieve the stress such shrinkage places on a large expanse of concrete, the material will crack. The same condition exists when we're talking about Portland cement plaster. Such cracks not only affect the aesthetics of a stucco finish, but represent a weakening in the overall system's effectiveness since the stucco protects and supplements the moisture resistance of the weather barrier membrane.

To minimize the number and severity of cracks–and even control where they will occur–we can break large expanses of stucco into smaller units, by placing control or expansion joints between them. Control/expansion joints allow the cement plaster to crack (imperceptibly) in a neat, straight line at the joint, rather than in some unplanned location in the middle of the stucco surface. Control and expansion joints also make the task of hand-applying stucco easier, as they allow time-critical application steps (e.g., floating of the brown coat) to be completed in one area before moving on to the next area.

You are much more likely to see control (and expansion joints) on commercial projects than residential ones. The easy explanation is that wall areas tend to be smaller in residential construction. Local conventions and practices would also suggest that homeowners would object to the way they look aesthetically (breaking up wall areas, as they do) and that they often introduce water intrusion issues. Nevertheless, the previously-mentioned ASTM C 1063 requires that control joints to be installed (in commercial OR residential construction) so that:

- Depending on the coarseness of the texture, no vertical surface (wall) may exceed 144 square feet unbroken.
- No horizontal surface (soffit, overhang, ceiling) exceeds 100 square feet unbroken.
- The distance between control joints should not exceed 18 feet in either direction.
- The length-to-width ratio of an area bounded by control and/or expansion joints should not exceed 2.5:1.
- Control joints in stucco over solid substrate walls (i.e., concrete or masonry) need be installed only at locations that control joints are installed in the solid substrate.

On walls and ceiling surfaces that use metal lath to anchor the stucco, you should divide the area into rectangular panels with a control joint. Do not extend the metal lath across these control joints. Instead, trim the lath on either side of the control joint, and attach the lath to the control joint with wire ties. Use metal that is corrosion resistant for control joints on exterior surfaces.

Table 2-1

Lath accessories

Zinc alloy is more corrosion-resistant, but is ductile and difficult to install true and straight. Galvanized steel (ASTM A653, G-60) performs well and A653, G-60 should be considered for buildings near the ocean.

 Casing Bead or Plaster Stop	Used as a cement plaster stop bead and provides a neat, exposed trim around windows and doors.
 Foundation Weep Screed	Used at the bottom of wall surfaces, behind the weather-resistant barrier, to allow water behind the stucco finish to drain down and away from the structure. Required by codes and published standards.
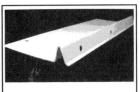 Foundation Weep Screed (Vinyl)	Used at the bottom of wall surfaces, behind the weather-resistant barrier, to allow water behind the stucco finish to drain down and away from the structure. Since a bottom screed isn't as likely to receive abuse as a corner bead would be, vinyl can be a suitable material for this accessory.
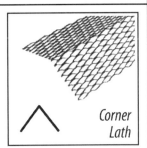 Corner Lath	Made from expanded metal mesh, Corner Lath is used on inside angles/corners. It is usually supplied in a greater-than-90-degree angle, to provide a tight fit when installed
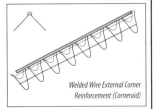 Welded Wire External Corner Reinforcement (Corneraid)	External corner reinforcements are required by code for reinforcing outside corners. More typically fabricated of welded wire than expanded metal lath. Available in bullnose configuration and plastic-nosed for exposed 90-degree corners.

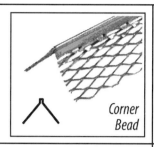 *Corner Bead*	Primarily for interior applications. Used for outside corners, the corner bead provides reinforcement close to the nose of the stucco bead. While available in zinc and galvanized steel (and even vinyl), only zinc or galvanized steel should be used in exterior applications.
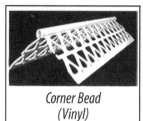 *Corner Bead (Vinyl)*	Interior applications. Used for outside corners, provides limited reinforcement close to the nose of the stucco bead. For greater durability, consider using the zinc version instead.
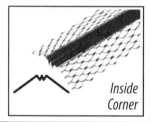 *Inside Corner*	Sometimes used in interior corners where two building sections meet, to allow for shifting/movement at the corner. Should be zinc for exterior applications.
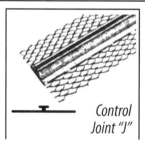 *Control Joint "J"*	Made from zinc (for exterior use), control joints relieve the stresses of expansion and contraction in large cement plaster surfaces. Where sound transmission and/or fire ratings are prime considerations, an adequate seal should be provided behind the joint.
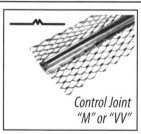 *Control Joint "M" or "VV"*	Relieves the stress of expansion and contraction in large plaster surfaces. While control joints are offered in both zinc and galvanized steel, zinc should be used in exterior applications.
 *2-Piece Expansion Joint Slip Joint*	Relieves the stresses of shear, expansion and contraction in large plaster areas. While they are offered in both zinc and galvanized steel, zinc should be used in exterior applications.

Soffit Screed (Vinyl Only)

2", 3", and 4" soffit vent screed provides a continuous one-piece vent for stucco applications in soffit areas, allowing attic ventilation. Beads along the edges act as casing beads, providing a screed surface for stucco application. Since a soffit screed isn't as likely to receive abuse as a corner bead would be, vinyl can be a suitable material for this accessory.

BEADS

Casing Beads (Plaster Stops): Install casing beads along the top of each wall and around each window and door opening. These beads support the lath and stucco, establish grounds and provide a proper key for the plaster.

Exterior Corner Reinforcements: Installed at outside corners for reinforcement and to help create true, even edges.

- Cut casing beads and corner reinforcements to size, using a metal snip or hacksaw. Trim other accessories to fit if necessary.
- Fasten the accessories to the concrete foundation or the concrete/masonry base with forced entry fasteners or hardened masonry nails or to the edge of the wood sheathing using galvanized steel nails.
- Be sure you install all beads level or plumb, and align sections/pieces properly.

GROUNDS / BASE SCREEDS / WEEP SCREEDS

A large assortment of base or weep screeds is available, for various special uses in stucco work.

Generally, the term ground or screed is used to refer to devices for establishing the proper thickness of the plaster membrane. (See the discussion of "rodding" on page 85.)

Accessories with an extended lip/embedment flange should be used to key stucco to the component.

- Grounds are narrow strips of metal or wood attached to the base adjacent to edges and openings in the plastered area. Door and window frames are frequently used for this purpose.
- A screed is similar to a casing bead; it provides a clean edge to the cement plaster and helps set the finished plaster's depth.
- A weep screed is a special kind of screed used along the bottom of a wall surface. The weep screed is installed behind the weather-resistant barrier. Any water that is able to penetrate the stucco surface is stopped by the water-resistant barrier, and it flows down and out over the nose.

INTEGRATION OF ACCESSORIES, LATH AND BARRIER

CONTROL AND EXPANSION JOINT ISSUES

CONTROL JOINT INTERSECTIONS

HOW-TO'S
To join abutting ends of trim accessories, you should either splice or lap them, and seal them with appropriate sealants.

Applicable reference ASTM C 1063, Sections 7.10.1.4, 7.11.4 - 7.11.4.4

- Water-resistive barrier should be continuous behind joint intersection.
- Vertical control joints take precedence over horizontal control joints; therefore, verticals should run continuous and horizontals abut.
- Flanges should be mitered for neat fit.
- Abutting horizontal control joints should be set off 1/8 inch to 1/4 inch from verticals for expansion relief.
- Seal all intersections.
- Protective tape should be left in place until after stucco installation.

HOW-TO'S
To join control joint intersections, you should either splice or lap them, and seal them with appropriate sealants.

CONTROL JOINT INTERSECTION AT FLOOR LINE EXPANSION JOINT

- The upper portion of the water-resistive barrier is installed over the flange of lower flashing for positive drainage over the exterior finish.
- The lower portion of water-resistive barrier is installed above and behind the upper water-resistive barrier.
- Horizontally-oriented expansion joints should not be interrupted by vertical control joints.
- An abutting vertical control joint should be spaced 1/8 inch to 1/4 inch from horizontal control joint for expansion relief.
- Use sealant on control joint ends.
- See also "Weeping Expansion Joint at Floor Line" below.

ABUTTING JOINTS AT CORNERS

- Mitered joint intersections should be sealed with elastomeric sealant to protect from moisture entry.
- The same procedure applies to inside corners as outside corners.

ABUTTING CONTROL JOINTS

- Control joints and other accessories come in standard lengths of 10 feet. As a result, any run longer than 10 feet will require that the accessories be butted together.
- To ensure positive drainage, overlap the components, by clenching the bottom control joint so that it nests under the upper control joint.
- As an alternative, clench a small piece of control joint to splice the components together.

- Remember that all butted ends are vulnerable to moisture intrusion, so it is always a good idea to seal these ends with elastomeric sealant.

CONTROL JOINT ABUTTING WEEP SCREED

- The vertical control joint should be cut at an angle to accommodate the cant of the horizontal weep screed.
- The control joint bead should be held back slightly (approx. 1/8 inch to 1/4 inch) to accommodate eventual expansion into weep.
- End joinery should be sealed with elastomeric sealant.

TERMINATION OF STUCCO AT DISSIMILAR MATERIAL

Applicable reference: ASTM C 1063 section 7.11.3.

- A gap is allowed between the casing bead and the brick interface, to allow for insertion of the backer rod and elastomeric sealant by others.
- Expansion joints take precedence over casing beads, meaning that a casing bead would be interrupted by an expansion joint accessory whenever the two intersect.
- Interface of accessories should be sealed with elastomeric sealant.

#40 EXPANSION JOINT AT FLOOR LINE

- The water-resistive barrier should be continuous and unbroken behind the expansion joint.
- A piece of adhesive-backed membrane is sometimes used behind the joint for enhanced moisture protection.
- Expansion joint should be positioned at the floor line, with the upper (female) half attached through perforations along framing's bottom track. The bottom (male) half is attached through perforations along the framing's top track.
- Metal lath should lap the flanges of the joint.
- Closed-cell backer rod and elastomeric sealant can be installed for best resistance to moisture issues.

CASING BEAD EXPANSION JOINT AT FLOOR LINE DEFLECTION JOINT

- The water-resistive barrier should be continuous and unbroken behind the expansion joint.
- A piece of adhesive-backed membrane is sometimes used behind the joint for enhanced moisture protection.
- The expansion joint should be positioned at the head of the wall, with the top casing bead attached along the floor or stationary track.
- The bottom casing bead should be attached along the moving portion of double track.
- Metal lath should lap the flanges of the casing beads.
- Install backer and sealant.

WEEPING EXPANSION JOINT AT FLOOR LINE

- Slight cant on bottom break-formed flashing provides positive drainage.
- Break-formed flashing can include ground with cleat to create a mechanical key for the stucco.
- Lap the water-resistive barrier to provide redundancy behind the joint.
- The top layer of the water-resistive barrier should lap the lower flashing to allow for weep.

Note: Back to back casing beads or a #40 joint may also be used if the weep function is not a criteria.

WINDOWS

CASING BEAD AROUND WINDOW PERIMETER

Applicable reference ASTM C 1063 section 7.11.3.

- Penetrating elements should be isolated from the stucco.
- Stucco should be terminated with a casing bead around the window's perimeter to provide a cavity for the insertion of backer rod and elastomeric sealant.
- Casing bead at corners should be continuous rather than cut to help prevent water intrusion.

CONTROL JOINT MEETING WINDOW HEAD CORNER

- Provide a gap between the control joint and the casing bead of approximately 1/8 inch to 1/4 inch, to allow for lengthwise expansion of control joint.
- The meeting intersection should be sealed with elastomeric sealant and tooled for positive drainage.

Note: Extension of the drip cap flashing captures the bottom end of the intersection, preventing or limiting the amount of moisture getting behind backer rod and sealant at jamb.

- Sealant should also be applied to close any gap between the drip cap and the casing bead at the jamb.
- As an alternative, you can eliminate the casing bead at the head of the window and simply terminate the control joint at the drip cap.
- For "ganged" window units, it is preferable to use a single, continuous drip cap flashing rather than installing multiple, separate pieces between units.

CONTROL JOINT AT WINDOW SILL CORNER

- Similar to the window head corner situation (above), the control joint should terminate into the casing bead installed around the perimeter of window.

- A slight gap should be allowed (1/8 inch to 1/4 inch) for lengthwise expansion of the control joint.
- Meeting intersection should be sealed with elastomeric sealant.
- Closed-cell backer rod and elastomeric sealant should be installed later at jambs and sill.

ROOF LINE

PARAPET AT MEMBRANE ROOF

- For redundancy, the membrane from the roof should continue over the parapet wall and behind the stucco termination.
- Coping/cap flashing should employ interlocking, watertight seams.
- A continuous cleat should hold the coping/cap flashing in place over stucco.
- Consider sealing the interface of the coping and stucco with elastomeric sealant, to prevent wind-driven rain from intruding to the interior.

PARAPET / WALL INTERSECTION

Applicable references: IBC 1405.3, IRC R703.8.

- Beams and parapets that intersect with high walls are susceptible to water intrusion issues.
- This can best be addressed by fabricating a "saddle" flashing that can be installed prior to the cap flashing/coping.
- The saddle flashing can be installed prior to installing the roofing membrane on the parapet or after.
- The saddle flashing should be installed over stucco for best results.

FASCIA / SOFFIT WEEP

- This provides a means of weeping incidental moisture at interface of soffit.
- The projection of the casing bead below the soffit plane breaks the surface tension and thereby provides a natural drip stop.
- A vent screed is necessary only if there is an unconditioned air space above the soffit.
- The water-resistive barrier should be positioned behind the flange of the casing bead at fascia to provide for positive drainage.

SOFFIT VENT / WEEP SCREED

- This is typically a manufactured component that integrates a vent with the weep function.
- The water-resistive barrier should wrap onto the flange of the component for positive drainage from fascia to daylight.

OTHER TYPICAL INSTALLATIONS

ABUTTING CASING BEADS

- For continuity of some abutting accessory joints, skilled lathers will often overlap or splice components together.
- The back flange of the top casing bead can be cut out so that the ground nests into the lower casing bead.
- This method is preferable to simply butting the casing bead, especially at a moisture-vulnerable area (such as a termination at a window).
- Another method involves the fabrication of a separate splicer that is nested into both abutting pieces.

OUTSIDE CORNER AID

Applicable reference ASTM C 1063, section 7.11.2.1

- Lath is wrapped around the corner from one side to the other and extended to the next support.
- This pattern is repeated on the opposing side.
- Corneraid is then installed over the lath.

OUTSIDE CORNER BEAD

- Nosing at outside corner acts as a screed for proper thickness of stucco.
- While this provides a sharp corner, it is prone to rusting and separation cracking along nosing edge.
- Corneraid is a better solution, although creating a straight and square corner requires practice.

INSIDE CORNER LATH

- Discontinuous lath installed at an inside corner will result in cracking at this location.
- Lath is often bent at a 90° angle and continued to the next support.
- In lieu of bending lath through an inside corner, corner lath is sometimes used to provide continuity.

INSTALLATION ISSUES AROUND PIPE AND VENT SLEEVES

- Ensure that the weather-resistive barrier is snug around all breaches created by ducts, pipes, etc.
- Weatherproof with sealant or adhesive membrane tape as appropriate.

PROTECTING AND ISOLATING DEVICES INSTALLED PRIOR TO STUCCO

- If vent covers, electric meters and other devices are installed prior to the stucco, it may be necessary to not only protect them from the stucco, but also provide flashing or other measures to preclude moisture intrusion problems down the road.

WEEP SCREEDS

- A 3½" attachment flange should be attached at or below foundation plate line on exterior stud walls.
- The weep screed should be placed a minimum 4" above the earth and 2" above paved areas.
- The water-resistant barrier should lap the attachment flange.
- Lath should cover and terminate on the attachment flange of the weep screed.
- The actual "weep" function is provided by the cant of the weep screed's inverted "V." The perforations on the inverted "V" of weep screed function more as a mechanical key for the stucco. (See the discussion of "Weep Screed" above.)

THE WEATHER-RESISTANT BARRIER

With all of this talk about water intrusion and sealants, it should be no surprise that stucco usually requires a weather-resistant backing. Stucco goes on wet, so we need to protect underlying building materials from the wet, fresh stucco. But stucco is also a mildly porous surface, so that even set stucco can transmit exterior moisture (rain) to the building's interior.

As we've seen, when applying stucco to most concrete or masonry bases, no weather-resistant barrier is necessary.

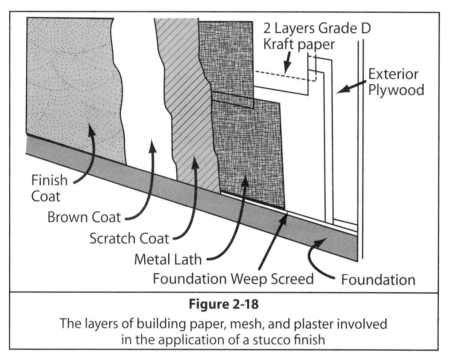

Figure 2-18
The layers of building paper, mesh, and plaster involved in the application of a stucco finish

Source: The Quikrete Companies, 2005.

FELT, KRAFT PAPER OR HOUSE WRAP?

Even if a building code didn't require it (yet nearly every building code does), unless your building was located in an extremely arid climate, you would still need a weather-resistant barrier beneath the exterior siding. That applies equally to stucco as to any other siding type, and the fact that stucco is applied wet is only part of the reason for the requirement.

Typically, a weather-resistant barrier is created by a layer of building paper or felt which is impregnated with asphalt. The asphalt in the paper/felt provides water resistance, but it also provides some measure of "seal" around nails and staples. Water can find paths through tears and holes that are created in this weaker material by rough handling, so be sure to repair any tears with construction tape. Felt is seldom used because of the difficulty in getting it to lay flat, particularly in cold weather, and the fact that most asphalt-saturated felt is a vapor retarder and prevents moisture from escaping.

⚠ CAUTION
Never apply stucco directly over wood, gypsum sheathing, or gypsum plaster. This also means that Portland cement plaster should never be applied directly to wood-based surfaces, such as plywood or OSB sheathing. For these bases, a water-resistant barrier is necessary.

⚠ CAUTION
Water-resistive barriers that have been degraded by excessive exposure or damaged by tearing should be replaced prior to the application of stucco.

Be careful: As in most code matters, it's up to your local inspector to determine whether the house wrap you're using is an "equivalent material." In most cases, you won't have a problem. But if you use a new or unfamiliar brand of housewrap, the inspector may ask you to provide the testing report for the product.

NOTE
If you're working in a locale that has adopted provisions like those found in the Model Energy Code (MEC), you may actually be required to use a housewrap.

The two most popular paper-based moisture barriers are "polyolefin fabric housewrap" and "Grade D Kraft" building paper. All model codes agree on the need for a weather-resistant barrier paper, and allow using one or both of these two choices behind stucco, brick, stone and other porous veneers. Grade D Kraft is stronger than felt, and is preferred over felt because, when wet, it allows water vapor to escape from the interior of structures better than felt. It is printed with a "minute" rating that indicates its ability to hold out water—generally 10 to 60 minutes.

In recent years, with the growing use of "plastic house wraps," some have begun to wonder which is better, and whether you should use both a house wrap and paper-based barrier material. Though No. 15 felt is usually cited as the standard for a weather-resistant barrier, nearly all building codes allow for the substitution of "equivalent" materials, opening the door for plastic housewraps.

Both No. 15 felt and Grade D Kraft provide good overall performance when it comes to water resistance; in fact, they actually perform better than some plastic wraps, but worse than others.

Plastic housewraps are engineered to prevent air infiltration and keep out liquid water, while allowing water vapor to escape from the building's interior. Grade D Kraft paper and plastic housewraps display these properties to one degree or another. No. 15 felt must be perforated to allow moisture vapor to escape. Paper-based materials actually have some advantages over plastic house wraps when it comes to protection against liquid water, however:

- It's true that paper-based products can lose strength and even rot when exposed to moisture over prolonged periods of time. But because felt and kraft paper are absorbent, they can actually wick water away from wall cavities–something plastic wraps just do not allow. (Plastic wraps allow water to escape only when it is in vapor form.)
- The permeability of paper-based barriers also varies, allowing more vapor through when they are wet than when they are dry. Most believe that Grade D Kraft performs better than felt in this regard.

These two factors make felt and paper more forgiving when it comes to some kinds of moisture intrusion. Finally, a paper-based barrier can be several hundred dollars cheaper than housewrap on an average home.

Alternatives to ASTM D 226 Felt that meet the criteria of IBC 1404.2 and IRC 703.2
• Fortifiber Jumbo Tex, Heavy Duty Jumbo Tex and Super Jumbo Tex Grade D Building Papers
• Dupont Tyvek Homewrap, StuccoWrap, and CommercialWrap
• BBA NonWovens Typar Housewrap (2003 IBC and IRC)
• Dow Chemical Weathermate Plus Housewrap

Alternatives to Grade D Building Paper that meet the criteria in IBC 2510.6
• Dupont Tyvek Homewrap, CommercialWrap, and StuccoWrap • Dow Chemical Weathermate Plus Housewrap • BBA NonWovens Typar Housewrap

HOW DOES THE BARRIER WORK?

No matter how solid and impressive it appears, siding–whether it's wood, brick, vinyl or stucco–does not create an impenetrable barrier against the elements. The truth is–whether water travels by wind, capillary attraction (absorption), gravity, or some combination of these forces–sooner or later water will find its way behind, around or through the siding layer. How can this happen?

During a rainstorm, wind can drive water against the building's exterior. Most of that water will be affected by gravity, and it will harmlessly drip or flow down the exterior surface. But stucco is porous to a certain extent. Combine that with the difference between the higher pressure on the exterior wall (created by the wind) and the lower air pressure inside a building, and you can see how water can literally be sucked into the interior, through any absorbent material or hole it finds. If there is no water-resistant barrier, water will be wicked toward the interior, where it can promote rot and other forms of water damage.

A weather-resistant barrier stops the travel of the water, and guides it safely down and away from the structure. Installing approved building kraft paper or housewrap on the exterior walls before siding material is applied is the first step to building this barrier system.

HOUSEWRAP OR PAPER... OR BOTH?

Most experts agree that if you get the flashing and accessory details right, install the building paper with proper overlaps, and avoid any gaps, breaks or tears in the paper layer, you will prevent nearly all of the moisture problems caused by wind-driven rain and snow. Approved housewraps and paper each provide an adequate secondary drainage plane. And both products are permeable enough to allow interior moisture to escape.

In those areas where a single-layer, paper-based barrier is allowed, the addition of a house wrap may provide an additional level of protection against air infiltration.

Many local codes require two layers of barrier over wood-based sheathing; in these instances, wrapping the structure with an approved house wrap, followed by building paper or paper-backed lath should be adequate. As always, check with your code or inspector to determine what your local area requires.

If there is no specific code requirement (which is rare), it really is going to be the choice of the builder or owner. And while it is tempting then to simply select the least expensive alternative, remember that the risk of rot, mold, mildew and worse should guide you not to shortchange the building in this area.

INSTALLING THE BARRIER

Water-resistive barriers should be installed with staples that do not protrude through the back side of the sheathing.

The water-resistive barrier should be installed flat and taut to the substrate surface.

The water-resistive barrier should remain unbroken behind all control and expansion joints.

In practice, building paper is installed soon after the sheathing. The paper itself must be properly layered, overlapped and taped where necessary to provide a clear drainage path.

But to be truly effective, the weather-resistant barrier must also be integrated with the flashing that is typically installed in later stages of a building project. So, as you study the discussion that follows, remember that the paper or felt needs to work with the flashing. And when the flashing is installed, care needs to be taken to preserve the weather-resistant barrier's integrity and design. Holes or tears created by the flashing installation should be taped or otherwise repaired to restore the integrity of the weather-resistant barrier.

START AT THE BOTTOM AND SHINGLE UP!

Since the primary objective of the weather-resistant barrier is guide liquid water down and away from the building, you'll use a shingle pattern:

- Start applying the first strips of paper or felt along the bottom edge of wall surfaces.
- Overlap adjacent strips side-to-side.
- Overlap the second and subsequent strips of paper or felt, over the preceding strips.
- This "shingle-ing" should be continuous over horizontal surfaces (e.g, soffits), reveals and architectural details; otherwise, you introduce the very real possibility of water intrusion and damage to the underlying structure.
- This shingle pattern should be followed on every layer of weather-resistant barrier material; that is, if your area requires two layers of paper/felt, the first layer should be applied using the shingle pattern, then the second layer should be applied using its own, separate shingle pattern.

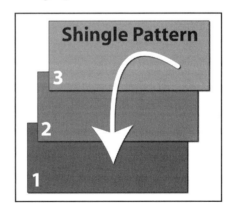

SAVE TIME WITH PAPER-BACKED LATH

It may seem we're getting ahead of ourselves here, talking about "lath," but in most applications, you can save time (and time is money) by using paper-backed metal lath. This product combines the required building paper pre-attached to expanded metal or wire mesh. The wire/lath sheet is typically offset from the underlying paper strip, and left free along the bottom and outside edges to allow for the separate overlap of paper and mesh.

ASTM C 1063 ("Standard Specification for Installation of Lathing and Furring to Receive Interior and Exterior Portland Cement-Based Plaster") requires that expanded metal lath shall be lapped ½" at the sides and 1" at the ends. The paper should be lapped a minimum of 2", shingle-style, although some jurisdictions require a 3" overlap.

Where paper-backed lath is used, the paper layer of each sheet should separately overlap the paper of the underlying sheet, and the lath of each sheet should separately overlap the lath of the underlying sheet, as shown in the figure below.

HOW-TO'S

Paper-backed lath is usually installed on a wood frame structure by first using using ordinary galvanized roofing nails or staples to tack it into place.

Once the sheet is in in place, self-furring nails are used every 6" O.C. along each framing member(stud) (do not nail into sheathing).

Typical paper-backed lath is intended primarily for vertical wall surfaces.

For horizontal surfaces such as ceilings, soffits or overhangs, you should use rib-lath.

Rib-lath is usually expanded metal mesh that features reinforced parallel ribs or grooves at regular intervals (2" or 4" are typical). The ribs provide it with rigidity, preventing the horizontal stucco surface from sagging.

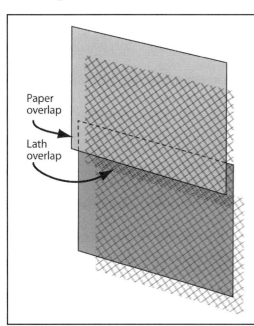

Figure 2-19
The offset of paper and lath/wire in paper-backed lath allows for the proper paper-to-paper and lath-to-lath overlap

- Attach the paper-backed lath, working from the bottom up. Position the first sheet so that the wire mesh rests in the stop bead/weep screed that runs along the base of the wall. You may need to trim the wire mesh. You can use a few galvanized roofing nails to tack the sheet in place, or staples.

- If using unfurred lath, the lath is attached to framing members (through the sheathing) with special self-furring nails that create a ¼" space between the lath and the paper. You need this space for good stucco adhesion. Use one nail per 6" O.C. per UBC, or 7" O.C. per ASTM and the International Codes.

👍 HOW-TO'S

Paper-backed lath is usually installed on a wood frame structure by first using using ordinary galvanized roofing nails or staples to tack it into place.

Once the sheet is in in place, self-furring nails are used every 6" O.C. along each framing member(stud) (do not nail into sheathing).

Typical paper-backed lath is intended primarily for vertical wall surfaces.

For horizontal surfaces such as ceilings, soffits or overhangs, you should use rib-lath.

Rib-lath is usually expanded metal mesh that features reinforced parallel ribs or grooves at regular intervals (2" or 4" are typical). The ribs provide it with rigidity, preventing the horizontal stucco surface from sagging.

- If using self-furring lath, attach the lath at furring points to framing members, being careful to maintain the self-furring spacing. Use one nail, staple, screw or wire tie at code-required spacing.
- The bottom edge of each course of paper should overlap the course below by two or three inches (depending on local code).
- The lath or wire mesh should overlap at least one full mesh.
- Stagger the start of paper-backed lath sheets so that vertical laps are staggered.
- The lath or wire mesh should extend ¾" under the corner reinforcements at any outside corners.
- Wire the mesh to any accessories (control joints, beads, etc.) with wire ties, at least once every 7 inches. (Accessories need only be secured to hold them in place during plastering, but 7 inches is a good rule of thumb.)
- Wire the overlaps in the mesh at least once every 9 inches between supports.

INSTALLING PAPER AND LATH SEPARATELY

- Tack or staple the paper in place, using galvanized staples or nails.
- Each horizontal course of backing should overlap the course below it by two inches minimum.
- If using unfurred lath, the lath is then attached to framing members (through sheathing) with special self-furring nails that create a ¼" space between the lath and the paper. You need this space for good stucco adhesion. Use one nail per 6" O.C. per UBC, or 7" O.C. for ASTM and International Codes.
- If using self-furring lath, attach the lath to framing members being careful to maintain the self-furring spacing. Use one nail per 7" O.C. per UBC, or 7" O.C. for ASTM and International Codes.
- If using wire mesh, each sheet should overlap adjoining sheets at least one full mesh.
- If you are using expanded metal lath, the lath should at least overlap at each side by ½", at ends by 1".
- Install wire mesh or lath sheets so that vertical laps are staggered and do not occur at jambs of openings.
- Where lath is not continues around the corner, the edges of abutting sheets should extend at least ¾" under the corner reinforcements at any outside corners.
- Wire the mesh to any accessories (control joints, beads, etc.) with wire ties, at least once every 7 inches.
- Wire the overlaps in the mesh/lath at least once every 9 inches between supports.

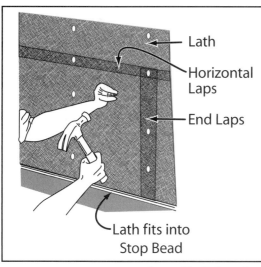

Figure 2-20
Nailing metal lath over weather-resistant barrier (Showing overlaps of lath)

For Weather Barrier
2" Horizontal Laps
6" End Laps
For Expanded Metal Lath
½" Horizontal Laps
1" End Laps

Source: The Quikrete Companies, 2005.

GENERAL INSTALLATION GUIDELINES

- Water-resistive barriers should be installed horizontally with the upper layer lapped over the lower layer a minimum of 2 inches.
- Where vertical joints occur there should be a minimum 6 inch overlap.
- The water-resistive barrier should be wrapped continuously across both internal and external corners.
- For wood-based sheathing, you will need two layers of water-resistive barrier. This is commonly accomplished with a double-laminate-layer roll of building paper or two consecutive installations of barrier material.

LATH

METAL LATH TO WOOD FRAMING

Applicable references: ASTM C 1063, Section 7.10.2.5 (1999); ASTM C 1002.

- Staples: 16 gage (0.062 in), 3/4 inch wide crown.
- Screws: No. 6, 0.136-inch shank. Type A, Pan or wafer head.
- Nails: 11 gage (min) roofing nail, with a 7/16-inch head. Or 6d common nails, driven in and clenched.
- Minimum ¾-inch fastener penetration into framing (stud) support.
- Minimum engagement of three strands of lath.
- Spacing along vertical supports of 7 inches O.C. in accordance with ASTM C 1063.

METAL LATH TO METAL FRAMING (USING SCREWS)

Applicable references: ASTM C 1063, Section 6.7.2. 7.10.3.3 (1999); ASTM C 954. C 1002.

- No. 6, 0.136-inch shank.
- 7/16-inch minimum diameter head. Pan or wafer head type.
- Self drilling and self tapping.
- Minimum 3/8-inch penetration into framing (stud) support.
- Head should engage a minimum of three strands of lath.
- Spacing along vertical supports of 7 inches O.C. in accordance with ASTM C 1063.

METAL LATH TO CONCRETE AND CONCRETE MASONRY

Applicable references: ASTM C 1063, Sections 7.10.4 and 7.10.5.

- Low velocity power or powder-actuated fasteners. (ITW Ramset Trakfast zinc coated or equivalent.) Shank diameter should be 0.145 inch minimum.
- Concrete stub nails: 3/4 inch long with 3/8 inch diameter heads.
- Minimum: Attach with six power or powder-actuated fasteners located at each corner of lath and at mid point along long edge of each side. Then stub-nail balance at 16 inches O.C. horizontally and 7 inches O.C. vertically.
- Better: Use power or powder-actuated fasteners entirely.

HOW-TO'S

Installation of lath to concrete and concrete masonry units with drill-and-drive fasteners, power or powder-actuated fasteners should be in accordance with manufacturers' recommendations.

NOTE

If there is any question about the effectiveness of the pull-out strength of drill-and-drive fasteners, power or powder-actuated fasteners into concrete or concrete masonry block, a sample testing is recommended.

DEFINITION

Self Furring Lath: Metal plastering bases having dimples, crimps, ribs or grooves designed to hold the plane of the back of the lath 6 to 10 mm (¼ to 3/8 inch) away from the plane of the substrate.

Table 2-2 Types of Lath
Woven Wire Lath is a product selectively used by the stucco industry. Because it resembles poultry netting it is sometimes referred to by the street names of "Chicken Wire" or "Stucco Netting." Over the years, its acceptance seems to have been relegated to installations of proprietary, one-coat stucco products. It is usually ordered by the dimension of its openings along with the gauge of the wire. For example 1-inch by 20-gauge.
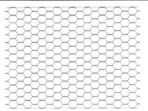 WOVEN WIRE FABRIC LATH Construction/Materials: 1½" x 17-gauge hexagonal wire mesh. Residential and light commercial use. This mesh is available flat or self-furred. When flat is used, self-furring fasteners should be used. A self-furring fastener comes with a ¼" fiber wad surrounding the shaft. Position the wad behind the lath and against the weather-resistant barrier/wall. Hook the lath over the nail or screw head and drive the fastener into the frame.

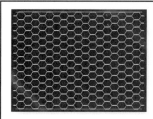	**PAPER-BACKED WOVEN WIRE FABRIC** Construction/Materials: 1½" x 17-gauge wire mesh, backed with Grade D Kraft paper. Residential and light commercial use. The mesh in this combination is almost always self-furring, so self-furring fasteners may or may not be required. A self-furring fastener comes with a ¼" fiber wad surrounding the shaft. Position the plug between the lath and the paper. Hook the lath over the nail or screw head and drive the fastener into the frame.

As the name implies, Welded Wire Lath is constructed of strands of wire that have been fused together by welding. It is distinguished by its heavier-gauge wire and 2-inch square grid pattern, and typically includes a paper backing. It's use is currently limited to California and other areas of the Southwest, primarily.

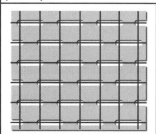	**WELDED WIRE LATH** Construction/Materials: 16-gauge galvanized steel wires fabricated into 2" x 2" squares openings, or 17-gauge wire lath with 1½" x 1½" openings. The 16-gauge product comes with factory applied, die-cut, absorbent paper backing. The paper does not provide a weather-resistant barrier; its purpose is to control/reduce the amount of cement plaster used, and relieve/reduce blowback by allowing pressure to escape to the back of the application surface. The 17-gauge product does not have factory-applied paper. Both are manufactured with furring crimps. Can be used in open frame or sheathed construction.

Diamond Mesh / Expanded Metal Lath

Expanded Metal Lath or Diamond Mesh Lath is made by cutting slits in steel sheets, then pulling to create the "diamond shaped" openings from which the mesh gets its name.

Expanded Lath comes in various weights, determined by the weight of the material per square yard; typical weights are 1.75 lbs., 2.5 lb. and 3.4 lb. The weight used depends on the application. Usually a galvanized version of this type of lath is preferred for tile-setting applications as well as stucco applications over unsheathed framing (open frame construction). This type of construction was typical for a number of years in the Southwestern United States, but is virtually unheard of today, as it doesn't meet the energy efficiency requirements in modern codes.

For applications of stucco over sheathing or solid surfaces, lath must be furred a minimum of 1/4". Self Furring Lath meets this requirement through a process which dimples the lath. Obviously, then, the lath needs to be positioned so that the dimples push inward against the solid surface.

Typically, local convention and practice for field-mixed three-coat stucco is to use 3.4 lb. self-furred galvanized expanded metal lath over sheathing and a previously installed water-resistive barrier. Another version of standard metal lath and furred lath is one that integrates a Grade D paper backing. So-called Paper Backed Lath installation can become a bit tricky in properly integrating the overlaps of adjacent sheets.

Ribbed Lath is a product that has roll formed parallel stiffeners and is often used in ceiling/soffit applications. Typically it can be fastened directly to the structural framing; however, more often it may be wire tied to gridwork that is suspended from the structure. Generally plaster keys quite well to this type of lath; however, a modification that is sometimes employed is preceding the rib lath with a Grade D paper backing, to minimize the amount of plaster that may flow behind the lath.

DEFINITION

Rib Lath: Metal plastering bases which have reinforced, parallel ribs or grooves at regular intervals. The ribs provide both furring (holding the lath off the back plane by a predetermined amount), but also rigidity. Rib lath is typically used on horizontal surfaces (ceilings, soffits, etc.) where framing spacing exceeds 16" O.C. but not more than 24".

> **DEFINITION**
>
> *Self Furring Lath: Metal plastering bases having dimples, crimps, ribs or grooves designed to hold the plane of the back of the lath 6 to 10 mm (¼ to 3/8 inch) away from the plane of the substrate.*

	DIAMOND MESH LATH Construction/Materials: A general-purpose lath which is slit and expanded from galvanized steel sheets, for application over CMU or concrete. It is easily formed around curved surfaces. Comes in both small (junior) and regular sizes; junior size for interior work or as strip lath reinforcements around openings (see Strip Lath). The rigid mesh provides improved keying for cement plaster when compared to woven or welded wire mesh. Used primarily in commercial construction.
Self-Furring Dimples	**SELF-FURRING DIAMOND MESH LATH (Dimple)** Construction/Materials: A general-purpose lath which is expanded from galvanized steel sheets for application over CMU or concrete. The rigid mesh provides improved keying for cement plaster. The dimples provide built-in furring, holding the lath the required distance (¼") from the base/barrier. Used primarily in commercial construction.
Self-Furring Groove	**SELF-FURRING DIAMOND MESH LATH (Groove or V-Groove)** Construction/Materials: A general-purpose lath which is formed from expanded galvanized steel. The rigid mesh provides improved keying for cement plaster. The horizontal groove (usually spaced 6" to allow for 6" O.C. attachment as required by most building codes) provides built-in furring, holding the lath the required distance (¼") from the base/barrier. Used primarily in commercial construction. Paper-backed version can be attached directly to studs in open frame construction.
OPENING *DOOR OR WINDOW*	**STRIP LATH** Construction/Materials: Expanded galvanized steel. Junior/small-sized diamond mesh lath pre-cut in 4" or 6" widths (usually supplied in 8 ft. lengths). Used at the corners of openings (i.e., doors and windows) or other areas where reinforcement is required to minimize cracks.
Plain Wire Fabric Lath	**PLAIN WIRE FABRIC LATH** Construction/Materials: Available in either 1½" x 1½" or 2" x 2" square or hexagonal grids, fabricated from 16- or 17-gauge galvanized wire, welded or woven. Used infrequently. Also called "stucco mesh." Wire is "ribbed" (furred) to provide the proper key.

STOPPING LATH AT CONTROL JOINTS?

A subject that is a matter of some industry discussion is the subject of cutting or stopping lath so that it is discontinuous at control joints.

- One position suggests that the bellows (the "M" profile in the control joint) must remain independent of the lath so that it can open and close freely with expansion and contraction of the stucco. The only way this can happen is to cut the lath with a power shears then position and wire tie the control joint over the cut lath.

- The other position states that whether you cut the lath or not, this whole question is moot since the lath is attached to the structure which negates any "independent" opening and closing. Moreover, stucco in its plastic state, squeezes in and behind the joint and in effect fuses the cut lath back together as it hardens and cures.

Most industry groups and bureaus prescribe to the second position. Until these controversies can be reasonably settled, it is suggested that all accessories be attached to the substrate as required by ASTM C 1063 section 7.11.1.1. In the event that the accessory must be attached in a position between supports, it should then be wire-tied to the lath. Whether the flange of the accessory is positioned behind the lath or on top of the lath then becomes inconsequential to the overall effectiveness of the installation, provided that the flanges of the component are embedded in the stucco.

A NOTE ABOUT SELF-FURRING METAL LATH

In most residential work over a typical, flat wall surface, woven wire fabric lath is not only the least expensive but an entirely satisfactory choice. In commercial or custom residential work, or wherever unusual surface shapes exist, the type of lath specified is frequently a heavier-gauge metal lath–often the expanded metal lath variety.

In California, the California Building Code says that: "Metal lath and wire fabric lath used as reinforcement for cement plaster shall be furred out away from vertical supports at least ¼ inch (6.4 mm). Self-furring lath meets furring requirement, except when installed over plywood sheathing or similar rigid backing. The use of self-furred lath is subject to a satisfactory jobsite demonstration for each project of the lath installation, with approval by project architect and the enforcement agency. EXCEPTION: Furring of expanded metal lath is not required on supports having a bearing surface width of 1-5/8 inches (41 mm) or less."

So using self-furring lath (typically expanded metal lath with self-furring grooves) in California, you'll be required to demonstrate the attachment techniques and fasteners you'll use before the inspector will provide approval.

If you are attaching self-furring lath, you need to follow the same rules regarding where and how you attach it. Care should be taken to maintain the self-furring spacing; that is, attach self-furring lath only at the dimples, ribs or grooves provided for this purpose.

- Attach the lath, working from the bottom up. Rest the first sheet of mesh in the foundation weep screed that runs along the base of the wall.
- Attach the self-furring lath to framing members, being careful to maintain the self-furring spacing. Use one nail per 6" O.C. per UBC, and 7" for ASTM and International Codes.
- If using wire mesh with a self-furring feature, each sheet should overlap adjoining sheets at least one full mesh. If you are using self-furring expanded metal lath, the lath should overlap at each side by ½", at ends by 1".
- Install wire mesh or lath sheets so that vertical laps are staggered.
- The wire mesh should extend ¾" into the corner reinforcements at any outside corners.
- Wire the lath to any accessories (control joints, beads, etc.) with wire ties, at least once every 9 inches between supports.
- Wire the overlaps in the lath at least once every 7 inches.

WOOD LATHS

Wood lath construction has largely been replaced with the systems described above; however, in a repair or renovation job, or in an historical restoration, you may encounter wood laths.

In some cases, wood laths (thin strips of wood, usually ¼" x 1") are attached over paper to provide a 3/8" "key" for the stucco. This is an older technique, requiring significantly more labor and more expensive materials; therefore, it is primarily used in restoration and repairs. It is very rare in exterior work, and is more often found in interior applications.

- First attach a series of vertical laths ("furring strips") to the underlying structure, over the paper/felt. The spacing you are able to use on the furring strips will depend on whether you are attaching them directly to studs or framing members, or to sheathing.
- Now attach horizontal lath slats (again ¼" x 1" wood) leaving a gap of no more than 3/8" between them. This gap provides the "bonding key" for the first coat of cement plaster.

EXTERIOR INSULATION AND FINISH SYSTEMS (EIFS)

The term "Exterior Insulation and Finish System" (EIFS) is used to refer to an exterior wall cladding system where rigid insulation boards are applied to the exterior of the wall sheathing. These insulation boards are then covered with a stucco-type exterior coat. EIFS finish coats come in a wide variety of colors and textures similar to those you'll find for stucco; however, EIFS textures cannot be built up as heavily as normal stucco finishes. Color retention in the finish coats is relatively good.

EIFS has been used in the United States since the early 1970s, after meeting considerable success in Europe. Early systems consisted of expanded polystyrene (EPS) bead board glued to the sheathing and covered with a modified cement base coat that included woven glass fiber reinforcement. This was followed by a textured/colored finish coat. This system has evolved over dozens of years, so that now there are a variety of "systems."

> **NOTE**
> Today, your research job is much easier: most EIFS manufacturers provide complete information (specifications, materials lists, construction details, tips, answers to frequently asked questions, etc.) on their websites. Chapter 11 provides a list of manufacturers and their website addresses.

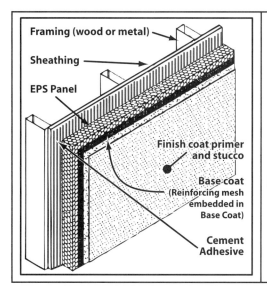

Figure 2-21
Layers in a typical EIFS installation.

NOTE: The adhesive ribs should be oriented vertically to facilitate drainage of intruding moisture.

Today, the insulation boards are typically closed expanded polystyrene (EPS), but in some cases extruded expanded polystyrene (XPS) panels are used. These insulation boards are attached to the underlying sheathing or building frame, using either special adhesive or mechanical fasteners.

WHY EIFS?

The reasons for EIFS' growing use and acceptance include the following:

- Energy-efficient – the insulating panels' ability to reduce thermal loads on the exterior building wall. (Note: The thermal performance of EIFS systems relies heavily on the thickness of the insulation panels used. The insulation should never be modified to less than ¾-inch thick.)
- Light weight – little or no additional load on the structure.

- Lower labor and material costs – faster installation than many other siding choices, reduced number of application coats, reduced amount of coat materials.
- Design flexibility – the ability to form or sculpt the panels into shapes and patterns to achieve different architectural effects. This makes the EIFS approach especially useful when a project involves using implants/plant-ons to achieve architectural features.

EIFS – WITH AN EMPHASIS ON "SYSTEM"

EIFS are *proprietary systems:*

- "Proprietary" because each manufacturer develops their own combinations of materials and application techniques and requirements.
- "Systems" because the components of one system should not be substituted in another system, nor should they be modified or "stretched" beyond what the manufacturer's literature recommends.

The design of the entire EIFS wall system should be specified, and construction detail drawings should make clear the installation all components in the design of the final construction. If necessary, the step-by-step method of constructing the wall should be set forth in the details. In general, any EIFS terminations, openings, joints, surface-mounted objects and special surface treatments should be detailed.

Most EIFS manufacturers can provide three-dimensional construction detail drawings for façade interfaces, architectural effects, and related conditions that are useful starting points for using EIFS in your project.

Your best guide to whether an EIFS is what your project calls for, and what steps are involved in its installation, will be found in the manufacturer's documentation. And remember, your local building authorities may have some strong opinions about using any of these systems, so be sure to check with them before committing yourself to an EIFS.

In this book, we'll briefly describe some of the more popular systems, in very general terms, but without a specific manufacturer's system or a specific project in mind. It's always preferable to follow the detailed information the EIFS manufacturer/supplier will provide.

A VARIETY OF MATERIALS

POLYMER BASED (PB)

One of the more popular EIFS installations is the "Polymer Based" (PB) system.

- EPS insulation panels are installed, attached to the frame/sheathing using special adhesive or mechanical fasteners.

- A relatively thin reinforced base coat–just 1/8-inch-thick (nominal)–is applied. ("Reinforced" refers to the glass fiber mesh reinforcement that is embedded in the polymerized cement basecoat.)
- This is followed by the finish/texture/color coat (as thin as 1/16-inch thick).

POLYMER MODIFIED (PM)

A second EIFS installation type is called the "Polymer Modified" (PM) system.

- XPS insulation panels are attached to the frame or sheathing using mechanical fasteners.
- A thicker reinforced base coat–3/16-inch- to 1/2-inch-thick (nominal)–is applied.
- This is then followed by the finish/texture/color coat.

DIFFERENT WALL SYSTEMS / DIFFERENT PURPOSES

If you are considering an EIFS option, you may find a manufacturer saying things like "high-performance barrier EIFS" or "wall drainage EIFS." These different terms result in part from the material and finish used in the insulation panel, but they also result from the characteristics of the entire "system"–the type and thickness of the base and finish coats, the water-resistance/permeability of the insulation panel materials, etc.

BARRIER EIFS SYSTEMS

Barrier EIFS Systems rely primarily on the base coat portion of the exterior skin to resist water.

- Non-EIFS portions of the exterior wall must also share the same weather-resistant layer as the EIFS portion (e.g., use the same base coat material as the EIFS) so that the water barrier layer is continuous.
- Alternatively, non-EIFS portions of the wall should be sealed and flashed to prevent water from migrating behind the EIFS portions and into the underlying base/structure.

WALL DRAINAGE EIFS SYSTEMS

Wall Drainage EIFS Systems are more like other siding/stucco installations:

- The insulation panels are installed over a weather-resistant barrier layer that serves as a secondary drainage plane.
- The weather-resistant barrier must be properly layered, flashed and integrated into all other wall components to prevent water from migrating into the underlying walls or interiors.

RULES OF THUMB FOR EIFS INSTALLATION

Because every EIFS manufacturer has developed its own "system" of materials that work together, and its own set of instructions for installation and maintenance, this book can provide only a general introduction to EIFS installation and maintenance. You should read, understand and follow manufacturers' instructions.

The following are only general rules of thumb that are typically a part of most popular EIFS systems.

- Stagger the panels/boards, so that vertical joints from adjoining rows do not align with one another.
- The joints between foam boards should not align with the joints for any underlying sheathing.
- Panel/board ends at corners should be true/flush with the adjoining wall surface.
- Panels/boards at corners (whether inside or outside corners) should lap in an alternate fashion.
- Ensure adequate flashing and/or sealant joints around all penetrations (windows, doors, pipes, conduits, etc.).
- Don't forget kick-out flashing at the ends of roof flashing that butts up against the EIFS wall.
- Notch EIFS panels at the corners of doors and windows. Avoid panel joints at the corners of such openings. Install diagonal mesh at these corners.
- "Backwrap" the EIFS mesh around the edges of EIFS at terminations and penetrations.
- In wood frame structures, provide expansion joints at each floor/story line.
- Terminate EIFS above grade.
- When decks are installed over EIFS, use proper flashing.
- Apply base and finish coats evenly and at specified thicknesses over entire surface, including inside and outside corners, and at reveals.
- Provide adequate slope on top-facing horizontal surfaces to allow for drainage.

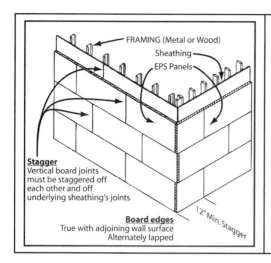

Figure 2-22
Installation pattern for EIFS panels

MAINTENANCE & REPAIR OF EIFS

A separate chapter in this book provides general information about stucco maintenance and repair. Most of the maintenance procedures are the same for EIFS, since the top, finish coat in EIFS is either stucco or very similar to stucco.

Otherwise, however, there are special methods for repairing damage to the EIFS and for maintaining its seals.

The base and finish coats used over EIFS foam panels are relatively thin–especially in Polymer Based (PB) systems–making the system susceptible to impact damage.

- Promptly repair any impact, hole or crack, and maintain proper seals at penetrations or terminations to ensure the lifespan of the EIFS and its continued ability to resist water.
- Repair methods for EIFS are simple and most manufacturers have published recommended methods for their particular systems.
- Some systems have exposed joint seals, and these must be maintained much the same as joints in other types of exterior sidings.

No construction technique is perfect, and EIFS systems are no exception. Problems with EIFS installations are primarily related to moisture intrusion. Remember, most EIFS systems rely on the base coat to provide water-resistance; however, moisture can migrate through window openings, at flashings and where holes and cracks in the EIFS allow water intrusion. Incorrect installation of the weather-resistant barrier in a Wall Drainage EIFS system, or improper or inadequate seals, flashing or coatings in a Barrier EIFS system can set the stage for sheathing and frame damage, mildew, etc.

This book can provide only a general introduction to EIFS installation and maintenance. You should read, understand and follow manufacturers' instructions. There have been significant problems with several EIFS systems

⚠ CAUTION
Particular care should be exercised when removing existing seal material from EIFS during maintenance or repair: don't damage the panels!

in the recent past; you would be wise to do plenty of research and familiarize yourself with all aspects of an EIFS system before choosing it.

ONE-COAT STUCCO SYSTEMS

These exterior Portland cement-based plaster systems – alternately referred to as "One-Coat Stucco" or "Cementitious Coating Systems" – were developed in the early 1990s as an answer to problems posed by the 1992 California energy regulations. These regulations could be met with 2 x 6 framed walls and conventional batt or blown-in insulation that many builders considered too costly.

Stucco manufacturers petitioned building code authorities to evaluate their proposed systems, constructed of 1"-thick rigid expanded polystyrene (EPS) board applied to the exterior of the 2 x 4 framing and secured to the studs with 1" x 20 gauge poultry netting stapled through the EPS into the studs. This substrate was then finished with a single 3/8" to ½" thick coat of a polymerized, fiber reinforced Portland cement plaster.

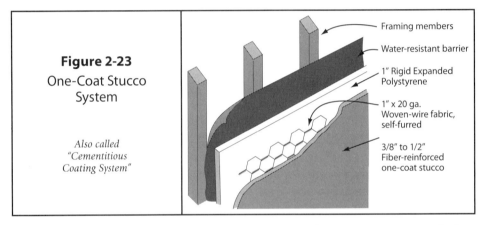

Figure 2-23
One-Coat Stucco System

Also called "Cementitious Coating System"

- Framing members
- Water-resistant barrier
- 1" Rigid Expanded Polystyrene
- 1" x 20 ga. Woven-wire fabric, self-furred
- 3/8" to 1/2" Fiber-reinforced one-coat stucco

Most (but not all) "one-coat system" manufacturers recommend the additional application of a finish material in the form of conventional mineral stucco, an acrylic stucco finish, tile or paint. Some of the manufacturers offer to add pigment to the basecoat to provide some color at lowest cost.

Due to a somewhat checkered history of job performance, the one-coat systems have, for the most part, been limited to use in entry level residential construction. The systems are most popular with builders looking for effective ways to cut construction time and costs. Many savings can be realized when measured against the alternative of using 2 x 6 wall construction. Actual time savings depend on the builder and his scheduling.

One-coat systems can be a very simple element in housing construction. The installation rules are straightforward and relatively easy to follow. A study conducted by the Building Industry Association of Orange County, California concluded that there appears to be less cracking in the one-coat systems when they are properly installed. Deviating from code-established minimum standards and/or the printed instructions of the one-coat system's evaluation report, however, is risky and must be avoided.

Since their first recognition by building codes as an acceptable substitute for conventional three-coat plaster application, the constituents of the one-coat stucco systems have changed. For example, some manufacturers no longer add resins to the formula, reducing the cost even further. The proponents have also expanded their recommendations to include application of their cementitious coatings over other insulating boards, gypsum sheathing, fiberboard and plywood.

Builders considering the use of a one-coat stucco system would be well advised to select one system, be sure that its code Evaluation Acceptance is current, and insist that the installation be done in strict compliance with the evaluation report. Do not intermix different manufacturers materials or adulterate the selected system's products.

The last page of every one-coat evaluation report contains an installation card to be signed by the coating manufacturers and the applicator certifying that the installation has been installed in compliance. A copy of the card must be presented to the building inspector.

CHAPTER 3

Plaster Mixes

DIFFERENT COATS = DIFFERENT MIXES

Tables at the end of this chapter present the materials to include in, and mix proportions for each coat in a stucco system.

- In a two- or three-coat stucco system, the basecoats (scratch and brown coats) use standard Portland cement as their base.
- The finish coat is typically a somewhat different mix–using white Portland cement as its basis when the final color is "bright," and gray Portland cement when the final color is on the "darker" side. Finish coat mixes also contain other additives for color and even weather protection.

For the most typical applications, there are now bagged stucco products designed to provide the proper basecoat for specific substrate conditions. Such products greatly simplify the mixing process (some include sand while others require the addition of sand), and help prevent costly mistakes.

The fact that there is a variety of mixes available, designed for different substrates tells us something: the plaster mix you use in your basecoats should be adjusted for the type of substrate to which they are applied. Table 3-1 presents some of the typical substrates you may encounter, and how the components of the mix should be proportioned accordingly.

PORTLAND CEMENT

The United States uses the specification prepared by the American Society for Testing and Materials: ASTM C-150 Standard Specification for Portland Cement, ASTM C 1328 Plastic Portland Cement, and ASTM C-91 Masonry Cement. A few other countries have also adopted one or more of these as their standard(s); however, there is no international uniformity in this area.

Though all Portland cement is basically the same, eight types of cement are manufactured to meet the different physical and chemical requirements of specific applications:

- Type I is a Portland cement suitable for most uses, and has historically been the most widely available material. (Type IA is a version of this cement used to make air-entrained concrete.)

"White" Portland cement is made of selected raw materials containing negligible amounts of minerals (iron oxide and magnesium oxide, primarily) that otherwise give cement its grey color.

White Portland cement is used wherever white or colored concrete or mortar is desired: stucco finish coats, precast curtain walls and facing panels, terrazzo surfaces, cement paint, tile grout, and decorative concrete.

- Type II is used for structures where the surrounding water or soil contains moderate amounts of sulfate, or where heat build-up is a concern. (Type IIA is a version of this cement that is used to make air-entrained concrete.)
- Increasingly, a blend of Type I and Type II is sold as general-purpose cement.
- Type III cement provides high strength at an early state, usually in a week or less. (Type IIIA is a version of this cement used to make air entrained concrete.)
- Type IV moderates the heat generated by hydration (the drying process), and is used for massive concrete structures such as dams.
- Type V is similar to Type II, in that it resists chemical attack by soil and water high in sulfates.

Cement for stucco usually should be Type I or Type II, a Type I/II blend. Two other hydraulic cements frequently used for stucco are ASTM C 1328 Plastic Portland Cement, and ASTM C 91 Masonry Cement.

- Type III (conforming to ASTM C 150) should be used if there are specific reasons; otherwise, Types I, II or I/II are preferred.
- Plasticizing agents in an amount not exceeding 12 percent by volume can be added to Type I or Type II cement, to make plastic cement.
- No lime or other plasticizer should be added on the job to plastic cement mix. When lime is used, it should conform to ASTM C 206, Type S.

SAND

The amount and kind of aggregates used greatly affect the quality of a plaster, the skills and methods required to apply the plaster, and the performance of the plaster in service. Proper grading of the aggregate (sand) is essential. To appreciate the role of well-graded sand in plaster, consider that almost 1 cu. ft. of damp loose sand is needed to make 1 cu. ft. of plaster: the sand makes up nearly the entire volume, whereas cement paste (a combination of cement, air and water) merely fills the voids between the sand particles.

Sand used in stucco mixes should be normal-weight, sharp sand (ASTM C 897 or C 144). Where weight or fireproofing is of prime importance, vermiculite or perlite (ASTM C 332) can be used. Sand should be washed clean, free of organic matter, clay, and loam, and well-graded. Too many fines (or dirty sand) increases the water demand and resultant shrinkage. Proof of any sand's acceptability is its service performance in the hardened plaster.

WATER

The amount and quality of water you use to when mixing the Portland cement plaster is critical in producing a workable product that also has long-term strength. Too much water reduces strength, while too little will make the plaster unworkable. Because Portland cement plaster must be both strong and workable, a careful balance in the water to cement ratio is required when making concrete.

- The water should be as clean and pure as possible – free of minerals, salt and other pollutants, in order to prevent side reactions from occurring which may weaken the concrete or otherwise interfere with the "hydration" process.

- Excessive water not only leads to an unworkable "soup," but it can also lead to the formation of capillary pores, which can seriously affect the material's strength and the durability.

- Water permeability increases exponentially when cement products are mixed with a water to cement ratio greater than 1:2, and durability decreases as permeability increase.

So, it's vitally important that you pay particular attention to and keep track of the water you're using in a mix. As always read, understand and apply specific manufacturer's/supplier's documentation in determining the amount of water you use. The following are provided only as guidelines:

- Add water using containers of known volumes (5-gallon pails, for example, with graduated markings), a water meter or a similar device that can keep track of the amount of each addition.

- Use a maximum 1:2 water to cement ratio when the cement will be exposed to freezing and thawing in a moist condition or to deicing chemicals. *1997 Uniform Building Code (Table 19-A-2)*.

- Use a maximum 1:2.22 water to cement ratio for cement where severe or very severe sulfate conditions exist. *1997 Uniform Building Code (Table 19-A-4)*.

PRECAUTIONS WHILE WORKING WITH CEMENT

Until fully wetted, the fine particles in dry Portland cement can present eye and breathing hazards. Once wetted, the cement is a highly alkaline solution (pH ~13), containing calcium, sodium and potassium hydroxides. Cement can cause chemical burns if contact is prolonged or if skin is not washed promptly.

- Wear gloves, safety goggles and a filter mask when working with the dry mix.

- Wear gloves and safety glasses/goggles when working with wet mix.

- All exposed skin areas should be washed after contact.

Once the cement surface dries/hydrates, it can be safely touched without gloves.

HOW-TO'S

The water to cement ratio is calculated by dividing the water in one cubic yard of the mix (in pounds) by the cement in the in the mix (in pounds).

So if one cubic yard of the mix has 235 pounds of water and 470 pounds of cement, the mix is a .50 (or 1:2) water to cement ratio.

If the mix lists the water in gallons, multiply the gallons by 8.33 to find how many pounds of water are contained in the mix.

MIXING SCRATCH & BROWN COATS

While you can mix cement plaster by hand, the limited amount of workable material that can be mixed by hand and the labor involved mean that hand-mixing makes sense only for stucco repair or very small stucco jobs. Cement mixing machines are inexpensive to rent, are widely available, and do a much better job of mixing than manual methods; therefore, machines should be used for all but the smallest stucco projects.

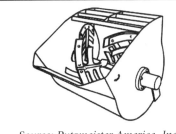

Figure 3-1
Paddle-type cement mixer

Source: Putzmeister America, Inc.

USING COMPLETE STUCCO MIXES

If you are using a bagged stucco mix, read, understand and follow the manufacturer's directions. If the manufacturer has not printed instructions on the bag, or if the mix is provided in bulk, follow this order:

- Start with the majority (60-70%) of the water you expect to use, in the mixer.
- Turn on the machine.
- Add all of the Portland cement and any admixture (such as coloring agent, plasticizer, etc.) Bonding agents (such as a weak PVA adhesive solution) will also aid bonding to existing masonry.
- Add more water slowly until you reach the desired consistency. Keep track of the amount of water you add at this point, so you can use that same total on subsequent batches.
- Once the desired consistency is reached, mix for an additional 3 to 5 minutes, until the material is uniform.

MIXING SEPARATE COMPONENTS ON SITE

If you are mixing materials on-site (and perhaps using a bagged Portland cement mix), follow this order:

- Start with the majority (60-70%) of the water you expect to use, in the mixer.
- Turn on the machine.
- Add approximately 50% of the sand.
- If you are using lime, add it to the machine now.
- Add all of the Portland cement and any admixture (such as coloring agent, plasticizer, alkaline fiber shorts, etc.)
- Add the remaining 50% of sand.

- Add more water slowly until you reach the desired consistency. Keep track of the amount of water you add at this point, so you can use that same total on subsequent batches.
- Once the desired consistency is reached, mix for an additional 3 to 5 minutes, until the material is uniform.

MIXING GUIDELINES

- Only mix as much cement plaster as you can apply in about 1½ hours. The actual time you will have to work with the mixed plaster will depend upon the temperature and other weather conditions; the mixed plaster will dry out quicker in hot, dry and/or windy conditions.
- Let experience be your best guide; start with small batches, and gradually increase the size of batches until you reach your comfort level.
- Any remaining mortar should be discarded; it cannot be "saved" and it should not be remixed.
- Water is the only ingredient in a cement plaster mix that should be adjusted, so it is better to err on the side of too little water, and adjust it upward.
- The best way to judge whether the right amount of water has been added is to apply a small patch and determine that it is workable and that it bonds well to the base and substrate. Once you have produced a satisfactory batch, use those same proportions for the balance of that workday, as long as weather conditions remain generally unchanged.
- The amount of water should not be adjusted much after the initial mixing period. Chemical reactions begin occurring within the cement once the water is originally added, and those components are used up; water added too much later will not produce the same result.
- Stucco should not be over mixed. If you are hand mixing, mix for only 10-15 minutes after adding the water.
- Machine mixing times vary by the speed of the motor and the efficiency of the machine's design; suffice it to say, machine mixing times (3 to 5 minutes after the last water addition, typically) are much shorter than hand mixing.
- If you are applying the stucco by machine (pump), you will need to use a cement mixer; hand mixing cannot provide an adequate and continuous supply of material to the pump. You may need to adjust the mixer's speed (faster) so that it completes a batch at around 3 minutes.
- Regardless of what mixing method you use, over-mixing can cause the stucco to set too quickly, which can lead to cracks and poor bonding.

DEFINITION

"Hydration" is a chemical reaction whereby the major compounds in cement form chemical bonds with water molecules, becoming "hydrates" or "hydration products."

Water is a key ingredient, which when mixed with cement, forms a paste that binds the aggregate together.

The water should be pure, to prevent side reactions that can weaken the product or otherwise interfere with the hydration process.

NOTE

"White" Portland cement is made of selected raw materials containing negligible amounts of minerals (iron oxide and magnesium oxide, primarily) that otherwise give cement its grey color.

White Portland cement is used wherever white or colored concrete or mortar is desired: stucco finish coats, precast curtain walls and facing panels, terrazzo surfaces, cement paint, tile grout, and decorative concrete.

MIXING THE FINISH COAT

The finish coat for stucco is typically a factory-produced material, consisting of gray or white Portland cement, fine aggregate, hydrated lime, pigment, and plasticizing agent, all mixed and sacked, ready for use. All that is needed is the addition of water on the job. Bagged products conform to building codes and are usually superior to a site-mixed finish coat because of their greater uniformity in properties and color.

USING A FINISH COAT MIX

If you are using a bagged mix for the finish coat, read, understand and follow the manufacturer's directions. If the manufacturer has not printed instructions on the bag, or if the mix is provided in bulk, follow this order:

- Start with the majority (60-70%) of the water you expect to use, in the mixer.
- Turn on the machine.
- Add liquid color or powdered pigments directly to the initial mixing water before you add the dry ingredients.
- Add all of the finish coat mix and any other admixture.
- Add more water slowly until you reach the desired consistency. Keep track of the amount of water you add at this point, so you can use that same total on subsequent batches.
- Once the desired consistency is reached, mix for an additional 3 to 5 minutes, until the material is uniform.

Carefully note the amounts/proportions you use, so that you can replicate the same mix (and, therefore, color) in each separate mix.

Table 3-1
Recommended Basecoat Plaster Mixes for Specific Substrates/Bases

Substrate/Base		Recommended Basecoat Plaster Mixes *Plaster Mix Symbols*	
		1st Coat	2nd Coat
Concrete; glazed and hard-fired brick	Low Absorption	C	C, CL, M or CM
		CM or MS	CM, MS or M
		P	P
Concrete masonry units (CMU), absorptive brick, unglazed tile	High Absorption	CL	CL
		M	M
		CM or MS	CM, MS or M
		P	P
Metal lath/reinforcement		C	C, CL, M, CM or MS
		CL	CL
		CM or MS	CM, M or MS
		M	M
		CP	CP or P
		P	P

Notes

Adapted from ASTM C 926.

Table 3-2
Basecoat Mix Proportions / Parts by Volume

Plaster Mix Symbols	Cementitious Materials					Volume of Aggregate (Sand) per Sum of Separate Volumes of Cementitious Materials	
	Portland Cement or Blended Cement	Plastic Cement	Masonry Cement		Lime	1st Coat	2nd Coat
			N	M or S			
C	1	–	–	–	0 - ¾	2½ - 4	3 - 5
CL	1	–	–	–	¾ - 1½	2½ - 4	3 - 5
M	–	–	1	–	–	2½ - 4	3 - 5
CM	1	–	1	–	–	2½ - 4	3 - 5
MS	–	–	–	1	–	2½ - 4	3 - 5
P	–	1	–	–	–	2½ - 4	3 - 5
CP	1	1	–	–	–	2½ - 4	3 - 5

Notes

Adapted from ASTM C 926.

The ranges shown for lime and sand content allow for adjustment of each mix to optimize the plaster's workability, depending on local sand material, installation and weight requirements. A higher lime content will generally support a higher sand content without loss of workability.

The second coat mix should use the same or greater proportion of sand as is used in the first coat.

Table 3-3
Finish Coat Proportions / Parts by Volume

Plaster Mix Symbols	Cementitious materials					Volume of Aggregate (Sand) per Sum of Separate Volumes of Cementitious Materials
	Portland Cement or Blended Cement	Plastic Cement	Masonry Cement		Lime	
			N	M or S		
F	1	–	–	–	¾ – 1½	1½ – 3
FL	1	–	–	–	1½ – 2	1½ – 3
FM	–	–	1	–	–	1½ – 3
FCM	1	–	1	–	–	1½ – 3
FMS	–	–	–	1	–	1½ – 3
FP	–	1	–	–	–	1½ – 3

Notes

Adapted from ASTM C 926.

Additional Portland cement is not required when Type S or Type M masonry cement is used.

For finish surfaces in areas subject to abrasion, specify mixes F, FP, FCM or FMS.

Better color will be achieved if a fine-grade, light-colored, washed sand is used as the aggregate for the finish coat. Sand should be graded No. 16 or finer.

Carefully weigh coloring compounds added to the mix, to ensure batch-to-batch consistency in color. It may be most convenient to pre-weigh and package colors prior to beginning the finish coat process.

CHAPTER 4

Applying Plaster

REQUIRED TOOLS & MATERIALS

- Stucco mixes or component materials (see chapter 3)
- Mixing box or cement mixer
- Mason's hoe
- Hawk
- Trowel
- Float
- Plaster rake
- Masonry brush
- Screed board
- Hammer

HOW MANY COATS?

You can apply stucco as a three-coat, two-coat, or one-coat system, depending on the type of stucco material used, the type of surface to which it will be applied, and the condition of that surface.

As we've discussed in other chapters, when it comes to smooth concrete walls, smooth masonry walls, and EIFS systems, a single, finish coat may be all that's needed. Otherwise, a stucco finish typically involves two or three coats of Portland cement plaster.

Table 4-1 (later in this chapter) provides general information on how thick each coat should be; consult your local codes and authorities for definitive information on this subject.

- The first layer is normally a minimum 3/8" (10 mm) thick and is referred to as a "scratch" layer because once it has been applied and has set up to ; the surface is "scratched" to provide a key for the second layer.
- The second layer, if used, is a leveling coat about 3/8" (10 mm) thick, and is sometimes referred to as a "brown coat." This coat must be rodded level (within ¼" over a 5-foot straightedge), and floated to densify and provide a suitable surface for the finish coat.

- The third or final layer is a thin covering about 1/8-inch (3 mm) thick, which may be colored and/or textured to give the final appearance.

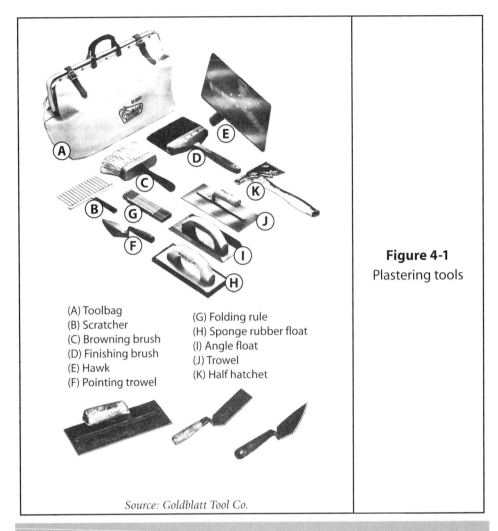

Figure 4-1
Plastering tools

(A) Toolbag
(B) Scratcher
(C) Browning brush
(D) Finishing brush
(E) Hawk
(F) Pointing trowel
(G) Folding rule
(H) Sponge rubber float
(I) Angle float
(J) Trowel
(K) Half hatchet

Source: Goldblatt Tool Co.

IS THE WEATHER RIGHT?

Temperature extremes make it difficult to properly cure stucco:

- On hot days, too much water is lost by evaporation.
- On the other end of the temperature spectrum, if the temperature approaches or falls below freezing, hydration slows to nearly a standstill. Under these conditions, the plaster ceases to gain strength and other desirable properties.

In general, the ambient (air) temperature during mixing, application and throughout the cure should not be allowed to fall below 40 degrees Fahrenheit (4 degrees Celsius).

ARE YOU READY?

- Cover the ground at the foot of the wall so that any dropped mortar can be easily removed. For this purpose, you can lay down some old planks or a tarpaulin, or spread some sand to keep the mortar off the ground.
- If you will be using scaffolding, it should be in place and secured according to all safety regulations. See Appendix D.

CONCRETE/MASONRY SUBSTRATE

- All control joints, other accessories and flashing for the entire project should be in place before beginning stucco application. If substrate conditions or design considerations dictate,, install self-furring metal lath using hardened masonry fasteners.
- Ensure that the surface is free of loose particles. If necessary, use a grinder or other mechanical means to provide a rough surface.
- Patch any significant holes or irregularities in the surface, and allow the patches to cure completely.
- On existing masonry walls, make certain that all mortar joints are solid; patch or remove any loose/brittle mortar. Re-point any joints that are significantly deeper than the rest. Allow any patches or joints to cure completely.
- It might be necessary to use a bonding agent prior to the first coat. Use an exterior-grade product, and apply it according to the manufacturer's directions.
- Wet the concrete or masonry substrate to a Saturated/Surface Dry condition: spray the entire wall until it is saturated, then allow its surface to dry slightly before beginning stucco application. This helps prevent the substrate from pulling too much moisture out of the cement plaster, which can cause cracking, loss of bond, and generally poor quality stucco work.

FRAMED/SHEATHED CONSTRUCTION

- Space ends and edges of wood-based sheathing panels 1/8" minimum, to allow for expansion.
- All control joints, other accessories, flashing, weather-resistant barrier and lath for the entire project should be installed before beginning stucco application.
- Tape or repair any holes or tears in the weather-resistant barrier.
- Make certain that the weather-resistant barrier layers (paper) have proper overlap and are shingled correctly.
- Make certain that the lath is attached at every framing member (stud), at 6" O.C. maximum, unless an International code applies, in which case 7" O.C. maximum spacing is allowed.
- Where lath sheets meet, make certain that a proper overlap is provided, and that the lath sheets are wired to each other every 9" maximum on side laps between supports.

- Make certain that the lath is properly wired to the casing beads, control joints, screeds, and other accessories, every 7" maximum.

HAND OR MACHINE APPLICATION?

Applying stucco by machine not only holds the promise of time and labor savings; with experience, it can produce a more consistent result.

Some plaster pumps feature integrated mixers (see Figure 4-4), so that the mixing/pumping process and the arrangement of equipment and materials is greatly simplified.

- Sprayed stucco is free of the lap and joint marks that characterize a surface that has been applied too slowly by hand.
- The machine-applied finish coat can be more consistent in texture, and it can deliver richer and more uniform colors.
- The machine-applied material contains slightly more water than hand-applied material, and slightly less air. (The impact of the pumped material hitting the base or previous coat drives much of the entrained air out.)

These few qualitative differences between the two methods are so slight as to be negligible. In addition, the performance/durability of the finish that each method can produce is nearly identical.

So, having put qualitative concerns aside, your decision to apply cement plaster by hand or by machine will likely be driven by a number of other factors, including the following:

- Your experience and familiarity with stucco materials and techniques.
- The size of the project.
- Accessibility to all areas of the site to be stucco-finished.
- Number of workers available.
- Overall expense.

SMALLER JOBS / FOR THE BEGINNER

For smaller jobs or for the beginner, hand application will typically be most appropriate choice.

- Hand application requires a limited outlay for tools, limits equipment rental to a cement mixer, and overall is a much more relaxed process.
- Whatever labor is available can be devoted to actual application, and no one needs experience or training in setting up, operating or maintaining a cement plaster pump.

MEDIUM-SIZED JOBS

For medium-sized (partial house) to larger jobs (full house), the decision whether to use hand or machine application will depend to some extent on the access to all surfaces to be finished.

- If large portions of the building are inaccessible or otherwise not amenable to machine application (due to problems with possible blow-back onto adjacent properties/structures, surfaces too distant from areas where the pump, mixer, materials and water can be set up, etc.), hand application may be the course to take. You can use wheelbarrows to bring mixed stucco from the mixer to the work area.
- If the building is mostly or completely accessible, machine application would be the best method.

LARGE JOBS

For large jobs–full houses, multiple houses, retail and commercial–machine application is nearly always the best choice.

- Overall labor cost is reduced by the much faster application times.
- Large projects typically justify more labor, and can support having workers devoted to mixing cement plaster and feeding and maintaining the pump.

MACHINE PLACEMENT AND SET-UP

Figure 4-2
Early plaster pump, 1930
Source:
Putzmeister America, Inc.

PLACING EQUIPMENT, MATERIALS, PIPE AND HOSE

A cement plaster pump must be "fed" with sufficient material to provide a steady stream of stucco to the nozzle.

- The cement mixer and pump, the mix materials, and the water source must all be located within very close proximity to one another.
- Map out each of the application "sessions" for the project in advance, remembering that full panels (areas bounded by beads and joints) of each coat must be completed in a single session.

- Rigid pipe should be connected to the pump outlet and used as far as possible before flexible rubber hose is used. (Rigid pipe creates lower resistance than flexible hose–not only because it provides a straighter run, but also because the metal material used in rigid pipe generates lower friction than the rubber used in flexible hoses.)
- Enough pipe and hose should be laid out to facilitate the application; however, the total length of pipe and hose used should not be excessive, to keep total resistance to a minimum.
- Make sure all couplings (including "quick-couplers") between pump and pipe, pipe and pipe, pipe and hose, pipe and whip line, and whip line and nozzle seal properly and do not leak. Leaks in these couplings are the leading cause of material obstructions developing in the line.

PUMP AND NOZZLE SET-UP

The whip line and nozzle are not connected to the hose until the sponge has been expelled from the hose.

Read, understand and follow all documentation, instructions and warnings provided by the cement plaster pump manufacturer or rental company. Differences among machines and new developments in equipment design mean that the following should be used only as general guidance.

Stucco application is typically a "wet gun" process, meaning that wet (already mixed) cement plaster is pumped to the nozzle. Air is added at the nozzle to propel the cement plaster to the target. This is also called "pneumatic application."

"Dry gun" techniques add water and air at the nozzle to a dry mix. Because the operator can vary the amount of air and water in a dry gun system (often without realizing it), product quality problems can result. In addition, a wet gun system produces half as much "rebound"–the waste created by sprayed concrete bouncing off the base surface and falling to the ground–as a dry gun system. That's why the stucco trade has almost universally adopted wet gun equipment as the application systems of choice.

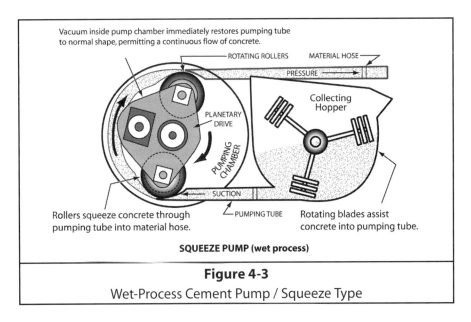

Figure 4-3
Wet-Process Cement Pump / Squeeze Type

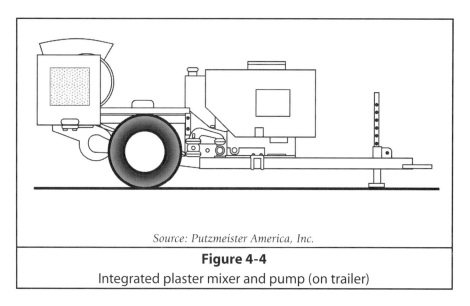

Source: Putzmeister America, Inc.

Figure 4-4
Integrated plaster mixer and pump (on trailer)

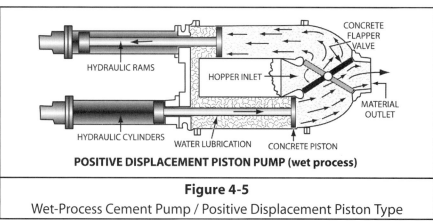

POSITIVE DISPLACEMENT PISTON PUMP (wet process)

Figure 4-5
Wet-Process Cement Pump / Positive Displacement Piston Type

- The pump is connected to the application nozzle, first using lengths of rigid pipe, followed by one or more sections of flexible rubber hose, and finally, with a lightweight, flexible "whip line." The whip line makes the application easier, as it facilitates aiming and movement along the surface.
- A separate air line provides the propulsive force necessary to throw the mix against the base surface with adequate force to form a proper key behind any lath, and to consolidate the sprayed material into a compact mass.
- The nozzle tip orifice is usually adjustable, and the nozzle tip itself can be changed, to provide different spray patterns and concentrations of material.
- Also at the nozzle are separate valves which allow the worker to control the air and material flows.

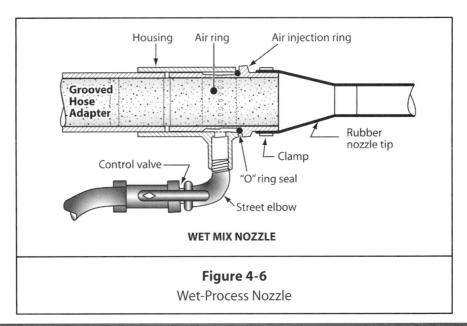

Figure 4-6
Wet-Process Nozzle

> ⚠ **CAUTION**
> *The whip line and nozzle are not connected to the hose until the sponge has been expelled from the hose.*

"PRIMING" THE PUMP

Immediately before each session of pumping cement plaster, the entire length of pipe/hose must be "wetted" and "lubricated" to ensure that the mixed material will flow easily and prevent damage to the pump and nozzle.

To "wet" and "lubricate" the pipe/hose line:

- Partially fill the pump's holding tank with water.
- Start the pump and move some of the water through the hose.
- Stop the pump and disconnect the pipe/hose at the pump outlet.
- Insert a wet sponge that is slightly larger than the pipe's diameter, and reattach the pipe/hose to the pump outlet.
- Start the pump again.
- When the holding tank is nearly empty, pour in a few gallons of a cement slurry (a watery mix of Portland cement and water without sand). This slurry will lubricate the interior surfaces of the pipe/hose
- When the sponge emerges from the end of the hose, stop the pump, and quickly attach the whip line and nozzle.
- Restart the pump and spray the remaining water and cement slurry into buckets or other waste containers.
- Just as the slurry is eliminated from the holding tank and nozzle, replace it with a proper mix of stucco material. Avoid delays here, as the cement slurry should not be allowed to dry or harden.
- Start the pump once the mix is ready, and still using waste containers, spray until "sand" appears at the nozzle.
- When the sprayed mixture appears uniform, you can begin application.

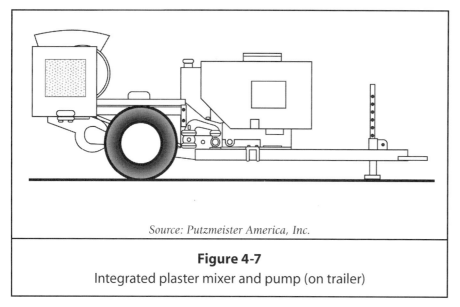

Source: Putzmeister America, Inc.

Figure 4-7
Integrated plaster mixer and pump (on trailer)

Most pumps these days have safety valves to prevent or minimize damage to the equipment due to obstructions. Occasionally, old or worn hoses can rupture before the safety valve kicks in, so it's just good practice to stop the pump immediately when material stops flowing through the line. The location of the obstruction should then be identified and removed.

CLEAN-UP AND MAINTENANCE

When work with the pump is finished for the day (or for any extended time during the day), the mixer, pump, nozzle and pipes/hoses should be cleaned.

- Remove the nozzle from the hose.
- Wash down the mixer bucket and paddles. To conserve water, the water used to wash down the mixer is released or poured into the pump holding tank.
- Additional water is used to wash down the pump holding tank.
- Start the pump, and pump the full amount of water through the hose, emptying the tank. Stop the pump.
- Uncouple the pipe at the pump, and insert the sponge that was used in the "wetting" procedure. Re-attach the pipe to the pump outlet.
- Add clean water to the pump holding tank. Run the pump until the sponge is expelled from the end of the hose.
- Attach the whip line and nozzle and run clean water through the nozzle.
- Wash off any cement plaster on the exterior surfaces of the mixer, pump, pipes, hoses, and nozzle.

APPLICATION PATTERNS

Regardless of how many coats–and regardless of whether you are applying a scratch coat or finish coat–there are certain guidelines you should follow when applying stucco.

First, establish and follow an application "pattern" that produces an evenly drying surface.

- Work within a limited area (e.g., one square yard/meter), and apply the stucco in that area to achieve the required thickness for that coat. Only then move onto an adjacent area.

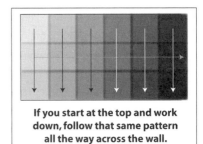

If you start at the top and work down, follow that same pattern all the way across the wall.

- In most cases, where the area bounded by control and expansion joints ("panel") is wider than it is tall, work from top-to-bottom or bottom-to-top in a column pattern (whatever you are comfortable with, or whatever the local practice is) before moving to the side.

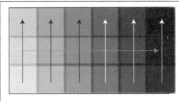

If you start at the bottom and work up, follow that same pattern all the way across the wall.

- Avoid using a row-by-row pattern: the time between adjacent squares (in one row to the next) on a typical wall can be too great, adversely affecting the workability and blend between edges, leading to cracks.

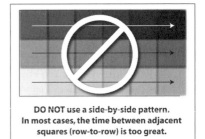

DO NOT use a side-by-side pattern. In most cases, the time between adjacent squares (row-to-row) is too great.

- The exception to this "rule" is when the panel is much taller than it is wide. In those situations, a row-by-row pattern may be preferable.

These patterns are clearly more critical for the hand-applied project, where application takes longer. But even in machine-applied stucco, a regular, well-considered pattern will produce better, more consistent results.

The hand-applicaton process is made easier by the "panels" that are established by the required control and expansion joints. You can easily complete work on a stucco coat in a single panel, following one of the recommended patterns above, before moving on to the next panel.

APPLICATION TECHNIQUES

THE FIRST (SCRATCH) COAT

When lath is being used, the first coat serves two purposes:

- To push cement plaster material into and behind the lath matrix, and
- To establish an top coating of material that sufficiently covers the lath so that the signature grooves of this coat can be "scratched" into it.

> **⚠ CAUTION**
> Whether you're applying the first coat over lath or concrete/masonry, avoid working the scratch coat too much, as this can interrupt the bond that begins to form almost immediately between the cement plaster and the base/lath.

HAND APPLICATION

In hand application of the first coat, you will make a few passes with a metal trowel, applying the scratch coat material over the lathed wall.

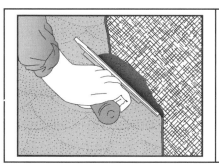

Figure 4-8
Applying the first coat

- Place some mix on a builders' hawk.
- Pick up enough plaster with a steel plasterer's trowel to cover about half of the trowel's surface, and spread it over the area.
- Follow one of the recommended application patterns (top-to-bottom or bottom-to-top).
- In the first pass, use just enough pressure to move material through and behind the lath while maintaining the "furring" space between the lath and backing. The stucco material should completely fill the space in a ¼- to ½-inch layer. In the case of a masonry substrate, the stucco should completely fill any joints, avoiding "ghosting" of the joints in the stucco material as much as possible.
- In the second pass, develop a uniform, required thickness of cement plaster of about ½" (13 mm).
- Complete the coat for a "panel" (the area bounded by required control or expansion joints) before moving on to the next panel.

MACHINE APPLICATION

- See the preceding discussion for general guidelines on preparing the pump and other equipment.

Read, understand and follow any documentation, instructions and warnings from the pump's manufacturer.

- In general, hold the nozzle sufficiently close to the surface (approximately 12") so that the scratch coat material moves in a concentrated fashion through and behind the lath and leaves behind, in a single pass, the required overall thickness of material.
- The angle of the nozzle with respect to the surface should be uniform, and each pass of material ("stroke") should overlap previous strokes by a uniform amount.
- Where strip lath, joints and other accessories are used (at control joints, along screeds, around windows and doors, at stop beads etc.), the nozzle should be brought within a 2-3" of the surface.
- Following machine application, some troweling may be required to bring the first coat to a roughly uniform thickness.

FOLLOWING BOTH HAND AND MACHINE APPLICATION

Allow the mortar to set up slightly to the point where it will hold a "scratch."

- Score the cement plaster horizontally, to a depth of about 1/8-inch (3 mm), about every ½".
- Do not rake the surface so hard that stucco is removed or falls away from the wall. The rake should move stucco material out of its path and deposit it to the sides of the tines.

You can use a stucco rake tool manufactured for this purpose (called, variously, a rake, a scratcher or scarifier), or you can make up a tool by driving a number of nails through a 2-to-3-foot length of lumber at ½" (13 mm) spacing. Make sure that all the nails project the same amount from the timber.

Figure 4-9
Rake tool for scratching stucco
(also called a "scratcher" or "scarifier")

Figure 4-10
Scratching the first coat with a site-made tool

CURE PERIOD

- Allow the scratch coat to set up for 24 to 36 hours. When applying stucco over open-frame wood construction, keep this coat continuously damp for at least 48 hours before applying the brown coat.
- Do not let the scratch coat dry out. The presence of moisture in the scratch coat is necessary for a good bond with the coat that follows.
- If necessary, keep the scratch coat damp by gently misting it with water. In very dry or hot conditions, you may also need to loosely wrap the structure with plastic sheeting to control the drying/curing time.

⚠ CAUTION
Do not use a high-pressure hose or nozzle, as excessive pressure can erode or dislodge the fresh cement plaster.

Figure 4-11
Mist the scratch coat to keep it moist

THE SECOND (BROWN) COAT

The brown coat serves two purposes, as well:
- To add to the thickness of the cement plaster surface (increasing the overall strength of the wall and its surface material), and
- To bring the surface to an even plane in preparation for the finish coat.

The brown coat can be applied as soon as the scratch coat has set enough to support the added weight. While this can be just a few hours in very hot or dry conditions, otherwise the brown coat should be kept damp for at least 48 hours then allowed to dry for five days.

If you won't be applying a third (finish) coat, the scratch coat and the brown coat should be applied more thickly thicker to create a 7/8-inch total thickness. The brown coat surface can then be textured. It is more typical, however, that a separate finish coat will be used, as the white Portland cement used in a finish coat product allows for coloring, and texture effects are easier to achieve.

HAND APPLICATION

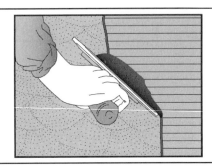

Figure 4-12
Applying second coat over scratch coat

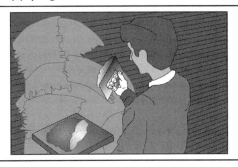

- Dampen the surface of the scratch coat a few minutes before beginning the brown coat application if it has been curing more than a few hours. Do not soak it and do not use a directed, high-pressure stream.
- Use a steel trowel to apply the brown coat material, pushing the material with enough pressure so that the new material completely fills the grooves in the scratch coat, and adds ¼" to 3/8" to the overall thickness of the combined stucco basecoat.
- Use one of the recommended application patterns (top-to-bottom or bottom-to-top) discussed earlier.
- Complete the coat for a "panel" (the area bounded by required control or expansion joints) before moving on to the next panel.
- As much as possible, try to make the brown coat as even in thickness as possible; this will reduce the amount of work necessary in rodding (see below).

MACHINE APPLICATION

- Dampen the surface of the scratch coat a few minutes before beginning the brown coat application if it has been curing more than a few hours. Do not soak it and do not use a directed, high-pressure stream.
- Using a technique similar to that used in applying the scratch coat, apply the brown coat material.
- The goal is to add the required thickness (usually ¼" to 3/8") to the overall basecoat.

- As much as possible, try to make the brown coat as even in thickness as possible; this will reduce the amount of work necessary in rodding (see below).
- Following machine application, some troweling may be required to bring the brown coat to a roughly uniform thickness prior to the next step: rodding.

FOLLOWING BOTH HAND AND MACHINE APPLICATION

After hand or machine application, the brown coat must be "rodded" and then "floated."

"Rodding" is the process of taking the overall stucco coating to a consistent thickness, and eliminating significant surface irregularities (lump, bumps and indentations). Because the finish coat is only 1/8" thick, it cannot not hide any bumps or lumps that are left in the underlying brown coat, so rodding is an absolutely necessary step.

- Rodding should take place just a few minutes after the brown coat is applied, while the brown coat material is easily worked.
- Run a "darby" over the surface of the brown coat, using stops, screeds, joints or wood slats along the top and bottom edges of the bounded area as guides, to bring the stucco to a consistent thickness.

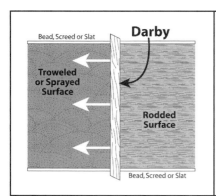

Figure 4-13
Rodding the Brown Coat Using a Darby

> **DEFINITION**
> A 'darby" can be metal or wood. The most important consideration is that it is long enough to span from the top to the bottom of the wall area bounded by joints, beads, screeds or wooden slats installed for the purpose of rodding. Professionals use metal darbies (typically aluminum), but a long, straight board–for example, a 1" by 4"–can serve the same purpose quite well.

"Floating" helps to consolidate the brown coat material and further improves the bond between the brown and scratch coats.

Floats come in a number of varieties: rubber, sponge rubber/foam, carpet, cork, etc. For floating the brown coat, use a sponge rubber or stiff foam float.

- Floating should occur only after the brown coat material has lost its watery sheen, but before it has become too rigid to be easily worked by the float.
- Move the float over the surface with a moderate amount of pressure; just enough to gently smooth the surface usually is sufficient.

CURE PERIOD

- Following the floating step, allow the brown coat to harden/cure for 24 to 36 hours.
- Keep it continuously damp by misting it with water through the 48-hour cure period. In very dry or hot conditions, you can loosely wrap the structure with plastic sheeting to control the drying/curing time.
- The brown coat can be allowed to dry out completely after the cure period has fully passed; however, it is absolutely critical that it is kept damp/misted throughout the cure period.
- If the brown coat has been allowed to dry out, it should be moistened just prior to application of the final/finish coat.

THE FINAL (FINISH/TEXTURE) COAT

Because the finish, texture or color coat involves a change in the mix (using white Portland cement instead of the standard, gray material) and texturing techniques, we'll cover this final coat in the next chapter.

HOW THICK?

Each successive coat builds the thickness (and ultimately the strength) of the stucco-finished surface.

The following table provides guidance on how thick each coat should be, in typical construction types. The figures in this table represent ranges taken from ASTM C 926 and a number of other building standards; therefore, it is especially important that you consult your local building codes and inspection authorities for the final word on requirements in your area.

Table 4-1
Stucco Coat Thicknesses (Nominal)[A]

		VERTICAL SURFACES (Walls)				
	METHOD	CEMENT PLASTER BASECOATS			STUCCO FINISH COAT	TOTAL THICKNESS
		1st Coat (Scratch)	2nd Coat (Brown)	Total of Basecoats		
Framed / Sheathed Construction[C]	Stucco Assembly with Lath[C,D,E]	3/8" - 1/2" 9 - 12.5 mm	3/8" 9 mm	3/4" - 7/8" 19 - 22 mm	1/8" 3 mm	3/4" - 7/8" 19 - 23 mm
Concrete Masonry Units (CMU)	Direct Application (2-coat)[E,G]	—	3/8" - 1/2" 9-12 mm	3/8" - 1/2" 9 - 12.5 mm	1/8" 3 mm	1/2" - 5/8" 12.5 - 16 mm
	Stucco Assembly with Lath[D]	3/8" 9 mm	1/4" - 3/8" 6 - 9 mm	5/8" - 3/4" 16 - 19 mm	1/8" 3 mm	3/4" 19 mm

	Method	1st Coat (Scratch)	2nd Coat (Brown)	Total of Basecoats	Stucco Finish Coat	Total Thickness
Concrete	Direct Application (finish coat only)[F]	–	–	–	1/16"-1/8" 2 - 3 mm	1/16"-1/8" 2 - 3 mm
	Direct Application (2 coat)[F,G]	–	1/4" 6 mm	1/4" - 3/8" 6-9 mm H	1/8" 3 mm	3/8" - 1/2" 9 -12.5 mm
	Stucco Assembly with Lath[D]	3/8" 9 mm	1/4" - 3/8" 6 - 9 mm	5/8" - 3/4" 16 - 19 mm	1/8" 3 mm	3/4" 19 mm

HORIZONTAL SURFACES (Soffits, Ceilings, Overhangs)[B]					
	METHOD	\multicolumn{3}{c}{CEMENT PLASTER BASECOATS}	STUCCO FINISH COAT	TOTAL THICKNESS
		1st Coat (Scratch)	2nd Coat (Brown)	Total of Basecoats		
Framed / Sheathed Construction[C]	Stucco Assembly with Lath[D,E]	1/4" - 3/8" 6 - 9 mm	1/4" - 3/8" 6 - 9 mm	5/8" - 3/4" 16 - 19 mm	1/8" 3 mm	3/4" 19 mm
Concrete	Direct Application (finish coat only)[F]	–	–	–	1/8" 3 mm	1/8" 3 mm
	Direct Application (2-coat)[F,G]	–	1/8" - 1/4" 3 - 6 mm H* & J*	1/8" - 1/4" 6 - 9 mm H* & J*	1/8" 3 mm	3/8" 9 mm J*
	Stucco Assembly with Lath[D]	1/4" - 3/8" 6 - 9 mm	1/4" - 3/8" 6 - 9 mm	3/4" 19 mm	1/8" 3 mm	7/8" 22 mm

NOTES for Table 4-1

A. For vertical surfaces (walls), the thickness of the plaster is determined by measuring from the face of the substrate to the outer surface of the stucco coat in question.

B. For horizontal surfaces (soffits, ceilings, etc.), the plaster thickness is measured from the back of expanded metal lath, usually rib lath (exclusive of the rib).

C. Sheathed framed construction is recommended rather than open-frame construction. The thickness of the cement plaster for these two types of construction is the same, however. Note: California requires sheathing on the first floor of multi-story frame structures.

D. In most locales, a basic stucco assembly should have a minimum thickness of 7/8 inches (22 mm), which can include a 1/8-inch (3 mm) stucco finish coat.

E If fire-resistance is required, the minimum thickness of cement plaster must be 7/8 inch (23 mm), which can include a 1/8-inch (3 mm) stucco finish coat. If the finish coat contains acrylic or other additives, the basecoat of cement plaster must be a minimum thickness of 7/8 inch (23 mm).

F. A liquid bonding agent should be applied to concrete before the application of a direct cement plaster basecoat or stucco finish coat (skim coat).

G. Cement plaster should not be applied direct to concrete or masonry substrates in thicknesses exceeding those shown here. If greater thickness is required, a stucco system with self-furring lath is required.

CHAPTER 5

Plaster Finishes

THE FINAL (FINISH/TEXTURE/COLOR) COAT

- As we've mentioned before, the final coat–also called the "texture coat" or "color coat"–is fairly thin: about 1/8" thick. The exception to this are specialty finishes (like "Travertine") that require a thicker finish coat. Also, some finish coats (as you will see in the pages that follow) require multiple finish coats.

- Most finish coats use a factory-blended mix, such as is made by California Stucco, Expo, Highland, La Habra, Merlex, Omega or others. Such mixes typically provide greater uniformity in color than job-mixed finish coats.

- If you do want to mix a custom color on the jobsite, remember that you will usually want to use a white Portland cement. (Standard Portland cement contains mineral oxides which give it a medium grey color, making it a less-than-satisfactory base for some color coats.) Add liquid color or powdered pigments directly to the initial mixing water before you add the dry ingredients.

- It is often best to produce one or more "sample panels" prior to applying the finish coat, to test the color, finalize the desired texture, and practice the finish strokes required to achieve it. These samples can be applied directly to scrap-quality plywood or gypsum wallboard without much preparation. A 4' x 8' panel size is usually required to repeat the texture pattern in full several times.

- As noted at the bottom of Table 3-3, truer color will be achieved if a fine-grade, light-colored, washed sand is used as the aggregate for the finish coat. Sand should be graded No. 16 or finer. The grade of sand can be varied within that range, to produce a desired texture effect.

- Carefully weigh coloring compounds added to the mix, to ensure batch-to-batch consistency in color. It may be most convenient to pre-weigh and package colors prior to beginning the finish coat application.

- Apply the finish coat (with all texturing) to an entire wall in a single session, to avoid differences in coloration or texture.

DEFINITION
A stucco finish coat applied directly to masonry or concrete is also called a "skim coat."

- Follow the application patterns (page 80) for the finish coat, as well. Complete a panel (the area bounded by beads, control and expansion joints) through all finish coat layers and texturing before moving on to the next panel. This is especially important for those texture effects that require multiple applications of finish coat material and significant time to texture.
- The finish coat can be *applied* using hand or machine application methods. Actual *texturing* (producing a desired pattern or texture on the finish coat) is always performed by hand.

PREPARING THE SURFACE

If the basecoat has been allowed to dry out fully, you should wet the surface slightly just before beginning the application of the finish coat material.

TEXTURE FINISH CHOICES

Different texture effects can be achieved by the use of multiple applications of finish coat material and troweling or scoring techniques.

On the pages that follow, you'll find photos and descriptions of a few of the textures that you can achieve with a stucco finish. As you can imagine, there is more art than science to texturing stucco, so let your creativity lead you. But if you don't feel particularly artistic or creative, don't worry: a typical stucco finish is rough and random. If you can keep that in mind, you shouldn't be afraid to jump right in.

Perhaps the most important thing to remember is that whatever texture you choose to employ, be sure it is something that you can achieve with consistency, and that the owner will want to live with it for years.

- Some of the samples either rely heavily on or benefit greatly from machine application (e.g., any finish requiring one or more dash coats). Others require hand finishing techniques. The notes for Hand and Machine attempt to give you a quick guide to the kind of labor involved.
- Some of the samples that follow might be fine on one building and not another. Things to consider when choosing a stucco texture include: neighboring buildings/local styles, simplicity/complexity of the building's architecture and detailing, etc.

When applying and texturing the finish coat, be consistent and work quickly.

- Carefully measure and consistently mix your finish coat materials and any color admixtures. Otherwise, noticeable batch-to-batch and day-to-day differences in color and texture can result.

- Many textures require a thin coat of the finish material to establish a color base, and then come back with additional finish material to create texture. Work quickly and finish a complete panel (areas bounded by beads, control and expansion joints) through all finish/texture steps before moving on to the next panel.

When applying finish coats by machine, the operator should know how to adjust the amount of air and spray pattern at the nozzle to produce desired effects (especially on those textures that employ dash coats).

TRAVERTINE

1. Apply the finish coat 1/4" to 3/8" thick over damp base. For this purpose, the finish coat material should be mixed slightly thicker (less water) than a typical finish coat.
2. For a realistic stone effect, two slightly different colors can be used and blended on the wall.
3. Rod and float the surface.
4. Layout horizontal joint lines with string attached to nails at either end of the wall surface, and either stamp joint lines with a stamp tool, or use another tool (masonry point, rake, etc.) Holding a small builder's square (or a piece of wood cut square) along the string or the horizontal scores in the surface, score vertical joint lines.
5. With a small stiff brush or other tool, pick out "chips" in the surface to simulate an uneven stone surface.
6. Water-trowel the surface smooth to remove ragged material, retaining most of the indentations.

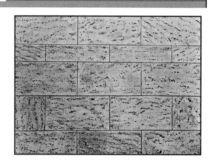

 Significant hand detailing/finishing.
 Can be used for coat application.
 Texture stamps and special tools. Thick finish coat.

ENGLISH

1. Apply a thin smooth coat of finish material, completely covering the basecoat.
2. Using a rounded trowel or similar tool, apply a second, thick coat in short, random strokes. This second texture coat will cover nearly all of the first finish coat.
3. The texture should be applied in a rough, random pattern, each stroke at angles to surrounding strokes.

 Mostly hand detailing/finishing.
 Can be used for first coat application.
 None.

SPANISH

1. Apply a thin smooth coat of finish material, completely covering the basecoat.
2. Using a trowel, apply a thin texture coat in a random pattern.
3. Adjacent strokes should overlap at angles to previous strokes.

 Hand — Mostly hand detailing/finishing.

 Machine — Can be used for first coat application.

 Special — None.

MONTEREY

1. Apply a thin smooth coat of finish material, completely covering the basecoat.
2. Apply a second coat of finish material in a random pattern. One way to achieve this is by throwing lumps of finish material onto the wall and spreading them with the trowel.
3. Adjacent strokes of random sizes and thicknesses should overlap previous strokes at random angles.

 Hand — Mostly hand detailing/finishing.

 Machine — Can be used for first coat application.

 Special — None.

KNOCKDOWN DASH

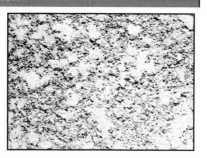

1. Apply a first dash coat (thin consistency; hand-thrown using a brush or machine-spattered onto the surface) to completely cover the basecoat for final color.
2. Apply a second dash coat of thicker, coarser consistency, for texture.
3. Allow some of first coat to show through the second.
4. After the second coat loses its watery sheen, run a trowel over the surface lightly to knock down the highest "peaks."

 Hand — Very little hand detailing/finishing.

 Machine — Best method for first and second dash coat applications.

 Special — Second dash coat is thicker/coarser.

MEDIUM DASH

1. Apply a first dash coat (thin consistency; hand-thrown using a brush or machine-spattered onto the surface) to completely cover the basecoat for final color.
2. Allow the first coat to dry.
3. Apply a second dash coat, covering nearly all of the first coat, for texture.

 Hand — No hand detailing or finishing.

 Machine — Best method for first and second dash coat applications.

 Special — None.

BROCADE

1. Apply a first dash coat (thin consistency; hand-thrown using a brush or machine-spattered onto the surface) to completely cover the basecoat for final color.

2. Apply a second dash coat of thicker consistency, covering nearly all of the first coat, for texture.

3. After the second coat loses its watery sheen, run a trowel over the surface lightly to knock down the highest "peaks."

Hand	Moderate finish work.
Machine	Best method for first and second dash coat applications.
Special	Second dash coat thicker.

LIGHT DASH

1. Apply a first dash coat (thin consistency; hand-thrown using a brush or machine-spattered onto the surface) to completely cover the basecoat for final color.

2. Apply a second dash coat of thinner consistency, covering nearly all of the first coat, for texture.

Hand	No hand detailing or finishing.
Machine	Best method for first and second dash coat applications.
Special	None.

TROWEL SWEEP

1. Apply a thin smooth coat of finish material, completely covering the basecoat.
2. With a small, short trowel, apply a second coat using fan-shaped strokes.
3. Successive strokes should overlap each other, leaving thin ridges where mortar flows over edge of trowel.

 Hand — Second coat requires hand work.

 Machine — Can be used for first color coat.

 Special — None.

RENAISSANCE

1. Trowel on finish coat using a back-and-forth motion. Alternatively, machine apply material and trowel to smooth the surface.
2. Brush the surface vertically with a coarse brush or broom (natural or synthetic bristles only; no metal wire brush), using light to moderate pressure. Keep the bristles free from mortar build-up by knocking the brush head against a hard, clean surface (e.g., a board) after each pass across the surface.
3. The stiffer the brush, the coarser the pattern.
4. Dampen the surface lightly to smooth the surface and remove any stray particles. A light pass with a float may also be used.

 Hand — Mostly hand detailing and finish work.

 Machine — Can be used for first/only coat application.

 Special — Brush with broom. Smooth with water and/or float.

MARBLECRETE

Marblecrete is an exposed aggregate finish, where small aggregate (about 3/8" diam.) (sometimes colored, sometimes natural) is dashed to a bedding coat. Since the application is somewhat different than most others, refer to manufacturer directions (also see Appendix C: Specifications, 10.10 Marblecrete). A "Marblecrete Hopper Gun" is available for applying the aggregate coat.

1. Apply the scratch and brown coats to proper thickness.
2. Straighten with rod and darby, leaving surface reasonably smooth, brushed in a horizontal direction only.
3. Apply a good quality bonding agent (Expo Bond, or its equivalent).
4. Apply aggregate matrix to bedding coat.
5. The aggregate may be tamped into the bedding coat for deeper embedment, or it can be left as applied.

 Typically some hand application and finish work.

 Pump can be used for base coat application. Special equipment available to apply aggregate.

 See Appendix C.

ACHIEVING FINISH TEXTURES

Smooth troweled finishes are not recommended for the following reasons:

- They are difficult to achieve.
- They are prone to excessive cracking.
- They often suffer from uneven shading from the troweling process.

If a smooth troweled finish is required, you can minimize cracking by embedding glass fiber mesh in 1/16" to 1/8" of cement adhesive (from an EIFS system) over the brown coat.

- Finishes on the smoother end of the spectrum can be created by running a steel float or trowel over the surface, occasionally plunging the tool into a bucket of water to keep it clean and minimize the amount of stucco it pulls away from the wall.
- Swirled textures are achieved by passing a float in a single, arc-shaped path over the finish material while it has a watery sheen. Do not work the surface a second time. Wooden floats (because of their softer surface) can produce better swirls than steel trowels.

Figure 5-1
Trowel and float techniques used for final texture effects

Source: HomeStore.com, 2005.

- A stippled texture can be achieved either by using a dash–splattering material on the surface using a machine applicator, or by flinging material from a brush. After the finish material has lost its watery sheen, hold a small, stiff hand broom (natural or synthetic bristles) at an angle to the wall, and pounce or pat the bristles with light to moderate force against the surface in a random pattern.

- Scratched textures start by first smoothing the second finish coat, and allowing it to dry (it should lose its watery sheen). Lightly draw a brush or broom (natural or synthetic bristles) across the surface. As the broom is drawn across the surface, you can use either a straight or wavy path, and a stiffer brush bristle will deliver a rougher finish. As you brush, frequently knock the brush against a clean, solid surface (e.g., a scrap board) to release any mortar build-up.

- Stamped or imprinted finishes typically employ metal or hard rubber "stamps." You can also make your own imprint device, but remember to allow for overlap and repeat of the pattern so that seams or gaps do not result. The final finish layer must sometimes be thicker than usual, to accommodate the depth of the imprint. Smooth the final finish layer and use the stamp or imprint device. If necessary, after the surface has dried for a few hours, you may want to come back with a float or brush to clean up and soften the edges of the imprints.

BUILT-IN COLOR

The stucco used in the scratch and brown coats dries to a drab, medium-gray color. To overcome this and provide interesting texture, most stucco finishes include the final coat we have been discussing. This finish coat stucco is formulated using white Portland cement and washed, finer-grade sand, which create a good, neutral (whitish) base for tinting.

- You can make bright white stucco by mixing white Portland cement, lime, and white silica sand for the finish coat.

In most cases, you will want to use a factory-blended finish coat mix, where the coloring is already added in proper proportions. This is not only the most convenient approach; it also will give you the best, most consistent results.

If you are looking for a color that is not already available in premix form, then coloring your own finish coat is an option.

Color pigments are available in powder and liquid forms.

- Regardless of the type you use, carefully measure the amounts you use as you mix the finish coat material, and mix each batch using the same proportions so you achieve a consistent color.
- Add liquid stucco and mortar color directly to the mixing water before you add it to the dry finish coat stucco mix.

TO PAINT OR NOT TO PAINT?

Two of the benefits of Portland cement plaster are its tough, natural texture and its long-lasting, built-in color; therefore, painting or otherwise coating plaster is often discouraged. Integrally colored stucco that is properly maintained (i.e., hosed down once a month to remove airborne dust and dirt), can be serviceable for 20 to 30 years; paint, on the other hand, must be reapplied every 7 to 8 years. That being said, it is occasionally desirable or even necessary to apply some form of surface protection or coating to repel water, fill fine cracks. to compensate for color variations–or to change the color.

Perhaps the best choice is to use cement-based paint, in an application called "fog coating." In fog coating, a fine mist of cement-based paint color is applied to the finished stucco surface. Since it is cement-based, this fog coat becomes integral to the stucco surface and will last nearly as long as an original stucco finish coat. (See Chapter 9 for more information on "Refinishing or Changing the Color" with a fog coat.)

LATEX PAINT

Latex paint is perhaps the least expensive option for freshening up a building's appearance or changing its color. Unfortunately and not surprisingly, latex paint provides the least amount of protection against moisture and limited durability. Paint can quickly decay, become chalky and peel. Stucco will not.

- Painting a stucco surface commits the property owner to regular re-painting.
- Worse, if the owner ever wants to return to a true stucco finish, the building must first be sandblasted to remove the paint.

The limited life of most painted finishes, and the sacrifice of stucco's easy maintenance, argues forcefully against using latex paint over stucco finishes.

- When used over patches or repair work–or over fresh stucco, for that matter–the stucco should be allowed to cure completely before applying latex paint.

ELASTOMERIC ACRYLIC / SILICONE COATINGS

Acrylic or silicone coatings are more durable than paint, and repel water better. These products must be applied much thicker than paint–typically a minimum of 15 dry mils. With this thicker application it is not surprising that these products can also fill fine cracks in the stucco better than thinner latex paints.

However, the water protection provided by some elastomeric coatings is indiscriminate: it prevents water from both entering and escaping from a building's interior. This "protection" may actually be too complete, even among coatings that claim they "breathe." By effectively sealing the cement plaster, they reduce its ability to dry out through evaporation. Such products set the stage for water condensation inside the wall structure, and can lead to mildew, rot, and other moisture-related damage.

Once coated with an acrylic or silicone coating material, the treated surface must be sandblasted before a true stucco finish can be restored.

- When used over patches or repair work–or over fresh stucco, for that matter–the stucco should be allowed to cure completely before applying the coating.

CHAPTER 6

Decorative Stucco

Many buildings today are being designed to look like the classic, Old World buildings found in Europe. These vintage structures were built of stone blocks that were trimmed by stonecutters and masons. The signature features of these buildings–decorative cornices, archways, keystones, medallions, eaves, quoins, etc.–were made of terra cotta cast from clay. Still other buildings are being designed to mimic buildings of the Federal style, with the style's signature columns, capitals, and cornices.

Now, all of these architectural details can be replicated by the use of "plant-ons" or "implants." A plant-on is basically a shape of formed expanded polystyrene (EPS) or similar material that is adhered to the plaster basecoat, and then covered (with the rest of the structure) by the finish coat of colored stucco.

- As can be seen on the pages that follow, window jambs, heads and sills can be made of flat or molded trims.
- The window head may be cornice shaped and the window sill is crown shaped.
- Roof eaves can be trimmed with cornices or reveals.
- Building corners can be finished to simulate lapped stone blocks called quoins.

Another way for decorating exterior plaster is to produce a surface that looks like stone masonry. This is done by simulated masonry joints. (See the Travertine finish in the previous chapter.) This texture can be applied by scratching horizontal and vertical lines into the finish coat, which then appears to be jointed masonry.

> **DEFINITION**
> *The terms "implant" and "plant-on" are interchangeable, referring to shapes (typically made from foam and wood, and covered with mesh or other materials). Once properly integrated with the water-resistant barrier and any affected flashing or accessory, and firmly attached to the building structure, these shapes are then finished with stucco.*

SHAPES ATTACHED TO STUCCO (PLANT-ONS OR IMPLANTS)

- To achieve a stone-block finish (Figure 6-1) and similar "architectural" effects, expanded polystyrene (EPS) foam shapes are attached to the stucco basecoat. Typically exterior and insulation finish stucco (EIFS) basecoat material or other material manufactured specifically for this purpose is used to make this attachment.

Figures 6-1
Building with Simulated Stone Block Finish

Source: Max Schwartz, 2005.

- The EPS foam shapes are finished as in a standard EIFS assembly (reinforcing mesh, basecoat and acrylic finish coat). The shapes can be assembled in the field, or pre-manufactured shapes may be used.
- The reinforcing mesh from the outside surface of an EPS foam shape should extend a minimum of 6 inches (150 mm) onto the surrounding stucco basecoat.
- An EIFS basecoat should then be applied over the extended mesh and feathered out onto the stucco basecoat.
- The stucco basecoat and EPS foam shape should then be finished with an acrylic finish coat.

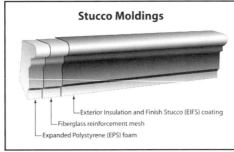

Figure 6-2
Stucco with Expanded Polystyrene (EPS) implant

Source: Stuc-O-Flex International, Inc.

When plant-ons are used, it is absolutely critical that their attachment to the structure does not compromise the integrity of the weather-resistant barrier. Furthermore, since some of these shapes create projections or reveals that can collect water, special flashing or water sealants are often needed.

The following figures illustrate just a few of the effects that can be achieved using plant-on and implants. Like EIFS, these are often proprietary systems that depend on a specific combination of materials (adhesives, reinforcing mesh, and coatings), so it's impossible to provide much greater detail here.

Read, understand and follow the documentation provided by manufacturers, distributors, etc.

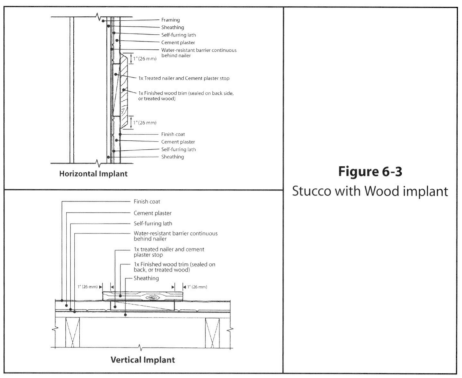

Figure 6-3
Stucco with Wood implant

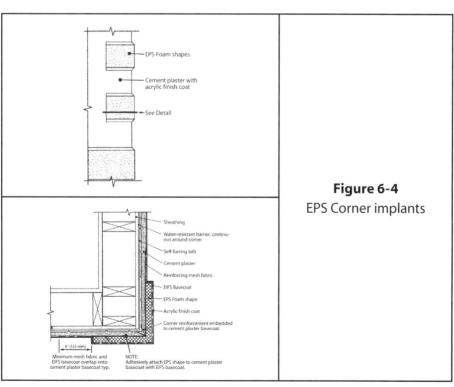

Figure 6-4
EPS Corner implants

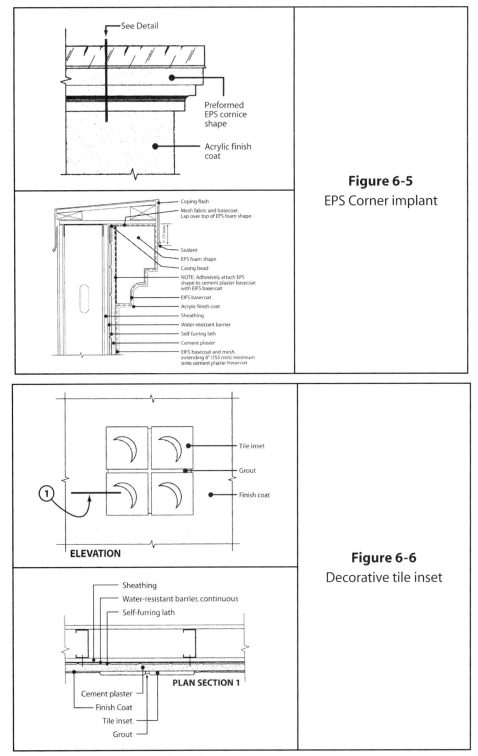

Figure 6-5 EPS Corner implant

Figure 6-6 Decorative tile inset

*Source for Figures 6-3 through 6-6:
Northwest Wall & Ceiling Bureau
Stucco Resource Guide (3rd Edition)*

CHAPTER 7

Properties of Plaster

STRUCTURAL STRENGTH

Wood frame buildings resist lateral (horizontal) forces from wind and earthquake by a combination of bracing and shear walls. Shear walls are vertical diaphragms consisting of plywood or particleboard panels, lath, and plaster. Both interior gypsum/wallboard and exterior cement plaster combine to provide resistance in shear walls to these lateral forces.

The strength of shear walls depends upon (a) the type and thickness of the materials used, and (b) nail spacing. You measure the strength of the material in pounds per lineal foot (plf) of panel. Plywood panels 5/16 inch thick have strengths that vary from 200 to 500 plf, depending on nailing.

Following are some typical shear strengths of plaster, for comparison:

- 7/8 in. Portland cement plaster with wire lath = 180 plf.
- 3/8 in. gypsum lath and ½ in. plaster = 100 plf.
- 5/8 in. gypsum wallboard = 100 to 150 plf.

Note that since the 1994 Northridge Earthquake (which is discussed later in this chapter), the exterior of most buildings in California over one-story in height are covered with plywood or particleboard and lath and cement plaster.

The 1996 Los Angeles County Building Code states in Chapter 25, that for the vertical diaphragm or shear wall, the studs must be 16 inches on center and the sills and plates must be adequately connected to the framing elements above and below to resist the horizontal forces. In addition, it requires that the height to width ratio should be 2:1 or 1-1/2:1, if the studs are blocked. However, for Seismic Zone 1, the height to width ratio is 1:1.

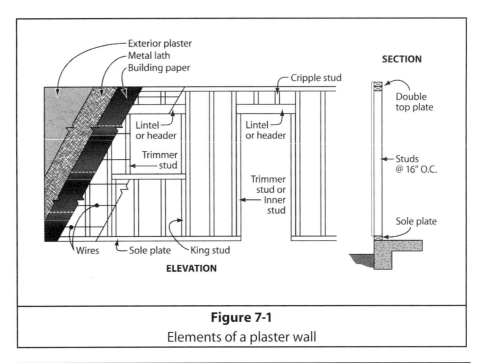

Figure 7-1
Elements of a plaster wall

Table 7-1
Height-to-width ratio of various types of sheathing

Type of Sheathing	Height/width ratio	Allowable shear (psi)
1" straight boards	2:1	50
2" straight boards	2:1	40
Diagonal boards	3.5:1	300
Lath and plaster	1:1	90
Metal lath and plaster	1:1	90
Gypsum wallboard	1:1	30
Plywood	2:1	75% of values in Table 23-I-J-1, 1994 Uniform Building Code

Source: 1997 Uniform Building Code

The following are excerpts from the 2002 City of Los Angeles Building Code and 2001 California Building Code, describing how plaster provides strength to a building for wind and earthquake.

SECTION 2513 - SHEAR-RESISTING CONSTRUCTION WITH WOOD FRAME

2513.1 General.

Cement plaster, gypsum lath and plaster, gypsum veneer base, gypsum sheathing hoard, and gypsum wallboard may be used on wood studs for vertical diaphragms if applied in accordance with this section. Shear-resisting values shall not exceed those set forth in Table 25-1. The effects of overturning on vertical diaphragms shall be investigated in accordance with Section 1605.2.2.

The shear values tabulated shall not be cumulative with the shear value of other materials applied to the same wall. The shear values may be additive when the identical materials applied as specified in this section are applied to both sides of the wall.

2513.2 Masonry and Concrete Construction.

Cement plaster, gypsum lath and plaster, gypsum veneer base, gypsum sheathing board, and gypsum wallboard shall not be used in vertical diaphragms to resist forces imposed by masonry or concrete construction.

2513.3 Wall Framing.

Framing for vertical diaphragms shall comply with Section 2320.11 for bearing walls, and studs shall not be spaced farther apart than 16 inches (406 mm) center to center.

Sills, plates and marginal studs shall be adequately connected to framing elements located above and below to resist all design forces.

2513.4 Height-to-length Ratio.

The maximum allowable height-to-length ratio for the construction in this section shall be 2 to 1. Wall sections having height-to-length ratios in excess of 1-1/2 to 1 shall be blocked. [LARUCP]

All shear walls designed to X resist lateral loads in Seismic Zone 4 shall have a maximum allowable height-to-length ratio of 1 to 1.

2513.5 Application

End joints of adjacent courses of gypsum lath, gypsum veneer base, gypsum sheathing board or gypsum wallboard sheets shall not occur over the same stud.

Where required in Table 25-1, blocking having the same cross-sectional dimensions as the studs shall be provided at all joints that are perpendicular to the studs.

The size and spacing of nails shall be as set forth in Table 25 -I. Nails shall not be spaced less than 3/8 inch (9.5 mm) from edges and ends of gypsum lath, gypsum veneer base, gypsum sheathing board and gypsum wallboard, or from sides of studs, blocking, and top and bottom plates.

2513.5.1 Gypsum lath

Gypsum lath shall be applied perpendicular to the studs. Maximum allowable shear values shall be as set forth in Table 25-1.

2513.5.2 Gypsum sheathing board

Four-foot-wide (1219 mm) pieces may be applied parallel or perpendicular to studs.

Two-foot-wide (610 mm) pieces shall be applied perpendicular to the studs. Maximum allowable shear values shall be as set forth in Table 25-1.

2513.5.3 Gypsum wallboard or veneer base

Gypsum wallboard or veneer base may be applied parallel or perpendicular to studs, except for certain fire ratings. Maximum allowable shear values shall be as set forth in Table 25-1.

TABLE 7-2
Allowable Shear for Wind or Seismic Forces, in Pounds per Foot, for Vertical Diaphragms of Lath and Plaster or Gypsum Board Frame Wall Assemblies[1]

TYPE OF MATERIAL	THICKNESS OF MATERIAL x 25.4 for mm x 304.8 for mm	WALL CONSTRUCTION	NAIL SPACING[2] MAXIMUM (inches) x 25.4 for mm	SHEAR VALUE x 14.6 for N/m	MINIMUM NAIL SIZES[3] x 25.4 for mm
1. Expanded metal, or woven wire lath and Portland cement plaster	7/8"	Unblocked	6	180	No. 11 gauge, 1-1/2" long, 7/16" head No. 16 gauge staple, 7/8" legs
2. Gypsum lath	3/8" lath and 1/2" plaster	Unblocked	5	100	No. 13 gauge, 1-1/8" long, 19/64" head, plasterboard blued nail
3. Gypsum sheathing board	1/2" x 2' x 8'	Unblocked	4	75	No. 11 gauge, 1-3/4" long, 7/16" head, diamond-point, galvanized
	1/2" x 4'	Blocked	4	175	
	1/2" x 4'	Unblocked	7	100	
4. Gypsum wallboard or veneer base	1/2"	Unblocked	7	100	5d cooler (0.086" dia., 1-5/8" long, 15/m" head) or wallboard (0.086" dia., 1-5/8" long, 9/32" head)
			4	125	
		Blocked	7	125	
			4	150	
	5/8"	Unblocked	7	115	6d cooler (0.092" dia., 1-7/8" long, 1/4" head) or wallboard (0.0915" dia., 1-7/8" long, 19/64" head)
			4	145	
		Blocked	7	145	
			4	175	
		Blocked Two ply	Base ply: 9 Face ply: 7	250	Base ply—6d cooler (0.092" dia., 1-7/8" long, 1/4" head) or wallboard (0.0915" dia., 1-7/8" long, 19/64" head) Face ply—8d cooler (0.113" dia., 2-3/8" long, 9/32" head) or wallboard (0.113" dia., 2-3/8" long, 3/8" head)

[1] These vertical diaphragms shall not be used to resist loads imposed by masonry or concrete construction. See Section 2513.2. Values shown are for short-term loading due to wind or due to seismic loading. Values shown must be reduced 25 percent for normal loading. The values shown in Items 2, 3 and 4 shall be reduced 50 percent for loading due to earthquake in Seismic Zones 3 and 4.

[2] Applies to nailing at all studs, top and bottom plates, and blocking.

[3] Alternate nails may be used if their dimensions are not less than the specified dimensions.

Source: Uniform Building Code.

PLASTER IN THE NORTHRIDGE EARTHQUAKE

The 1994 earthquake in Northridge, California was a wake-up call for plaster on wood-frame construction. The Wood Frame Subcommittee of the City of Los Angeles Department of Building and Safety and the Structural Engineers Association of Southern California (SEASC) conducted an extensive investigation, parts of which are described below.

Figure 7-2
Failed Stucco-on-Frame Apartment Complex

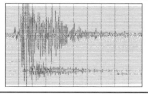

Figure 7-3
Seismograph record:
January 17, 1994
Northridge Earthquake

The investigation found that Portland cement plaster (stucco) and drywall shear walls performed poorly in the many multistory buildings, apartments, and condominiums during the Northridge Earthquake.

Figure 7-4
Failure of a 2-Story Apartment Complex in Northridge, California
Stucco on Frame Construction

Here are some of the types of framing failures that resulted from inadequate framing:

- Some narrow walls rotated and failed because the building had wood diagonal corner let-in bracing instead of plywood sheathing.
- Some walls bowed out because the 1 x 4 diagonal braces were not let-in or adequately nailed to the studs.
- Unsheathed walls rotated because they lacked let-in braces.
- Exterior lath and plaster separated from studs because the lath was not nailed to the studs with enough nails, or the nails used were too short.
- Brick chimneys fell down because they were not adequately anchored to the ceiling joists.
- Interior gypsum board shear walls failed because of inadequate nailing and edge blocking.
- Connections between the wood roof deck and masonry walls failed because of inadequate anchorage and nailing into the ledger.
- Rotational stresses occurred in asymmetrical buildings, causing portions of the building to separate due to rotation around shear walls of differing stiffness.
- Interior gypsum board walls acted like unintended shear walls and failed.

Because of these failures, the code now requires the following rules:

- Do not install plywood shear walls in line with gypsum board shear walls.
- Plywood shear walls failed because of inadequate or undersized nailing, or because nails were driven too close to the panel edge.
- Major structural failures occurred because there was a lack of connectors for continuous tension ties across intersecting framing members for roof and floor diaphragms.
- Due to vertical acceleration, the heavy HVAC equipment on roofs shifted. This over-stressed and stretched their bolted connectors.

After evaluating the damage, the Los Angeles City Building and Safety Department made the following changes in earthquake design of shear walls in wood-frame buildings:

- Reduce the shear value of Portland cement plaster from 180 plf to 90 plf.
- Reduce the shear value of all gypsum sheathing boards to 30 plf, although the 1997 UBC allows 100 to 175 plf for gypsum boards of various thicknesses and nailing.
- Limit the height to depth ratio (h/d) of gypsum wallboard shear walls from 2:1 to 1:1.
- Reduce the h/d ratio of plywood shear walls from 3:1 to 2:1.

- Prohibit use of Portland cement plaster, gypsum board sheathing, or gypsum wallboard shear walls to carry shear loads on the ground floor of every multilevel building.
- Allow only the use of plywood.
- Use only common nails.
- Use 3x members with a shear value of at least 300 plf for sill or sole plates and studs supporting panel edges.
- Increase the distance from the panel edges to the nails from 3/8 to 1/2 inch.
- Sheath cripple walls between the foundation and first-floor framing with plywood.
- Provide adequate anchor bolts for all mudsills.
- Provide adequate hold-downs for all shear walls.
- Use at least l/2-inch- diameter anchor bolts or bolts for hold-downs embedded at least 7 inches into the concrete.
- Install a minimum of two anchor bolts per piece of mudsill with one bolt located within 12 inches of the end of each piece.

THERMAL PROPERTIES

INSULATING QUALITIES

The thermal properties of building walls are very important in reducing heating and cooling costs. Plaster is an important component of the thermal resistance of a wall assembly.

Some of the terms that describe these properties are as follows:

- Thermal conductivity (K) of a material is described as:

    K = Btu/ (h) (sq ft) (inch thickness) (degrees F)

- Conductance (C) refers to the heat transfer through a material of any stated thickness, and is equal to:

    C = Btu/ (h) (sq ft) (degrees F)

- Resistance (R) refers to heat flow and transfer, and is equal to:

    R = 1/K or 1/C

- Heat loss (U) through a structure (a wall, floor, or roof) is described as:

    U = Btu/ (h) (sq ft) (degrees F)

- Conversion to the metric system:

    U value = 5.678 W / (sq meter) (degree C)

- The U value of any structural assembly, like a wall, is the reciprocal of the sum of the R-values for each of the components, and is:

    $U = 1/\sum R$

In addition to the R-values of materials, the surface of the structural assembly has specific resistance values. Here are some comparative thermal values:

Material	K	C	R (t/C)
Stucco, 1"	5.00	5.00	0.20
Wood siding, ¾"	0.80	1.06	0.93
Glass, ¼"	6.00		0.043

Source: Watson, "Construction Materials, and Processes," 2nd Edition McGraw-Hill Book Co.

TABLE 7-3
U Values of Several Wall Assemblies

Wall construction	U value
FRAME WALL	
Inside: metal lath and plaster	0.26
Outside: wood sheathing, building paper, wood siding	
Inside: 2-inch blanket insulation, metal lath and plaster	0.10
Outside: wood sheathing, building paper, and wood siding or 4-inch brick veneer	
BRICK WALL	
8-inch solid brick	
Inside: damp-proofed and plastered	0.49
Inside: 1 x 2 in furring, ½-inch insulation board, plaster	0.23
Inside: 2 X 2 in furring, 2-inch blanket insulation, vapor barrier, metal lath and plaster	0.10
10-inch, 2-inch cavity	
Inside: plaster	0.29
Inside: 2 X 2 in furring, 2-inch blanket insulation, vapor seal, metal lath and plaster	0.09
10-inch, 2-inch cavity, 2-inch insulation in cavity	
Inside: plaster	0.12
Metal curtain walls, with 2-inch insulation (average)	0.15-0.35
WINDOWS	
Single	1.13
Double	0.55
Triple	0.36

Source: Smith, Materials of Construction, McGraw-Hill Book Co.

FIRE-RESISTIVE PROPERTIES

Portland cement plaster and gypsum plaster provides fire-resistance to a building frame and thereby establishes its Type of Construction.

You classify the types of construction by the fire-resistive properties of the major elements of a building. These include the structural frame, walls, ceilings, floors, and roofs. Type I Construction is the most fire-resistive while Type V Construction is the least fire-resistant.

The terms that describe how a building will react to a fire are combustible, non-combustible, fire-resistive, and not fire-resistive. The words protected and non-protected may be used in place of the last two terms.

For example, a wood frame structure is combustible, while a steel frame one is non-combustible. A coat of plaster over a wood or steel beam provides some degree of fire-resistive protection. You base the degree of protection on the time required for a fire to weaken the member.

Typically, fire-resistance may be one, two, three, or four-hour protection. Various thicknesses of coatings produce different degrees of fire-resistance. Fireproof coatings may be of Portland cement, gypsum plaster, or similar mineral materials. Gypsum plaster is far more fire-resistive than Portland cement or lime plasters of equal thickness.

In summary, you classify buildings into five general types of construction: Types I, II, III, IV, and V. The codes further subdivide buildings into Fire-Resistive (F.R.), One-Hour (1-hr), Heavy Timber (H.T.), and No fire protection (N). You base these on whether the structural frame is combustible or non-combustible and how the frame is protected against fire:

TYPE I - F.R. — FIRE-RESISTIVE CONSTRUCTION

- This is the fireproof construction. Frames are non-combustible with three-hour fire-resistive protection. In some areas, you call these buildings as Class A Buildings.

TYPE II - 1-HOUR — FIRE-RESISTIVE, 1-HOUR PROTECTED CONSTRUCTION

- These are the next levels of fire resistant construction. Frames are made of a non-combustible material and have two-hour, one-hour, or no fire-resistive protection.

TYPE III - 1-HOUR, OR NON-PROTECTED, CONSTRUCTION

- Frames are combustible with one-hour or no fire resistive protection.

TYPE IV - H.T. (HEAVY TIMBER) CONSTRUCTION

- Frames are combustible, but made of heavy timbers, which are slow to burn.

TYPE V - 1-HOUR

- Combustible construction with one-hour fire resistive protection.

TYPE V - N

- Buildings with a combustible frame – or wood – without any fire protections. These are the most vulnerable to fire.

Table 7-4 A shows the basic allowable area, in square feet, for one-story buildings with Group B-2 and R-1 occupancies for different types of construction. A building area is the area within the exterior walls of a building exclusive of vent shafts and cores.

Table 7-4
Basic Allowable Floor Area for Buildings One Story in Height[1]
In Square Feet

	TYPES OF CONSTRUCTION								
	I	II			III		IV	V	
OCCUPANCY	F.R.	F.R.	1-HOUR	N	1-HOUR	N	H.T.	1-HOUR	N
A-1	Unlimited	29,900	Not Permitted						
A-2-2.1[2]	Unlimited	29,900	13,500	Not Permitted	13,500	Not Permitted	13,500	10,500	Not Permitted
A-3-4[2]	Unlimited	29,900	13,500	9,100	13,500	9,100	13,500	10,500	6,000
B-1-2-3[3]	Unlimited	39,900	18,000	12,000	18,000	12,000	18,000	14,000	8,000
B-4	Unlimited	59,900	27,000	18,000	27000	18,000	27,000	21,000	12,000
E-1-2-3	Unlimited	45,200	20,200	13,500	20,200	13,500	20,200	15,700	9,100
H-1	15,000	12,400	5,600	3,700	Not Permitted				
H-2[4]	15,000	12,400	5,600	3,700	5,600	3,700	5,600	4,400	2,500
H-3-4-5[4]	Unlimited	24,800	11,200	7,500	11,200	7,500	11,200	8,800	5,100
H-6-7	Unlimited	39,000	18,000	12,000	18,000	12,000	18,000	14,000	8,000
I-1.1-1.2-2	Unlimited	15,100	6,800	Not Permitted[8]	6,800	Not Permitted	6,800	5,200	Not Permitted
I-3	Unlimited	15,100	Not Permitted[5]						
M[6]	See Chapter 11, Uniform Building Code								
R-1	Unlimited	29,900	13,500	9,100[7]	13,500	9,100[7]	13,500	10,500	6,000[7]
R-3	Unlimited								

[1] For multistory buildings, see Section 505 (b), Uniform Building Code.
[2] For limitations and exceptions, see Section 602, Uniform Building Code.
[3] For open parking garages, see Section 709, Uniform Building Code.
[4] See Section 903, Uniform Building Code.
[5] See Section 1002 (b), Uniform Building Code.
[6] For agricultural buildings, see also Appendix Chapter 11, Uniform Building Code.
[7] For limitations and exceptions, see Section 1202 (b), Uniform Building Code.
[8] In hospitals and nursing homes, see Section 1002 (a), Uniform Building Code, for exception.
N-No requirement for fire resistance   F.R.-Fire resistive   H.T.-Heavy timber

Source: Uniform Building Code, 1991, ICBO.

Table 7-5 (below) shows the allowable height of buildings, in feet, for Group B-2 and R-l occupancies for different types of construction.

Table 7-5
Maximum Height of Buildings
In Feet and Stories

	TYPES OF CONSTRUCTION								
	I	II			III		IV	V	
	F.R.	F.R.	1-HOUR	N	1-HOUR	N	H.T.	1-HOUR	N
	MAXIMUM HEIGHT IN FEET								
	Unlimited	160	65	55	65	55	65	50	40
	MAXIMUM HEIGHT IN STORIES								
A-1	Unlimited	4	Not Permitted						
A-2-2.1	Unlimited	4	2	Not Permitted	2	Not Permitted	2	2	Not Permitted
A-3-4[1]	Unlimited	12	2	1	2	1	2	2	1
B-1-2-3[2]	Unlimited	12	4	2	4	2	4	3	2
B-4	Unlimited	12	4	2	4	2	4	3	2
E[3]	Unlimited	4	2	1	2	1	2	2	1
H-1[4]	1	1	1	1	Not Permitted				
H-2[4]	Unlimited	2	1	1	1	1	1	1	1
H-3-4-5[4]	Unlimited	5	2	1	2	1	2	2	1
H-6-7	3	3	3	2	3	2	3	3	1
I-1.1[5]-1.2	Unlimited	3	1	Not Permitted	1	Not Permitted	1	1	Not Permitted
I-2	Unlimited	3	2						
I-3	Unlimited	2	Not Permitted						
M[6]	See Chapter 11, Uniform Building Code								
R-1	Unlimited	12	4	2[8]	4	2[8]	4	3	2[8]
R-3	Unlimited	3	3	3	3	3	3	3	3

[1]For limitations and exceptions, see Section 602 (a), Uniform Building Code.
[2]For open parking garages, see Section 709, Uniform Building Code.
[3]See Section 802 (c), Uniform Building Code.
[4]See Section 902, Uniform Building Code.
[5]See Section 1002 (a), Uniform Building Code, for exception to number of stories in hospitals and nursing homes.
[6]See Section 1002 (b), Uniform Building Code.
[7]For agricultural buildings, see also Appendix Chapter 11, Uniform Building Code.
[8]For limitations and exceptions, see Section 1202 (b), Uniform Building Code.
N—No requirement for fire resistance   F.R.—Fire resistive   H.T.—Heavy timber

Source: Uniform Building Code, 1991, ICBO.

It is obvious that the allowable area and height of a building increases with the more fire-resistant types of construction. For example, a six-story apartment building must have Type I or II fire resistive construction.

A one-story apartment building with Type V 1-hour construction can have a basic area of 10,500 square feet. For multiple stories, you may increase the total allowable area to 200 percent of the basic allowable area, or 21,000 square feet. You can increase this even more by planning wider yards around the building or by adding automatic fire sprinklers.

For property located in a relatively low-density area, the code may permit a larger building area.

In summary, the conditions that set the maximum allowable area and height for a building are:

- type of construction
- group occupancy
- yards
- sprinklers
- fire zone
- number of stories

Table 7-6
FIREPROOFING COLUMNS
Using Lath and Plaster, Gypsum Wallboard and/or Vermiculite Concrete

Mark	Nominal Fire Resistance	Description
A	1 Hour	7/8" Portland Cement Plaster over Metal Lath tied to 3/4" Vertical Channels with #18 Gauge Tie Wire.
B	2 Hour	(Two Layers w/Air Space) 7/8" Portland Cement Plaster over Metal Lath tied to 3/4" Vertical Channel, 7/8" Portland Cement Plaster over Metal Lath tied to 3/4" Vertical Channels.
C	4 Hour	1" Vermiculite Concrete over Paper Backed Wire Fabric wrapped directly around Column with additional 2" x 2" Number 16/16 Gauge Wire over 3/4" Furring Channel and 1" Vermiculite Concrete outer surface.
D	4 Hour	1-1/2" Vermiculite Gypsum Plaster over Metal Lath wrapped around Column and furred 1-1/4" from Column Flanges. Plaster pushed through to Flanges.
E	2 Hour	1" Vermiculite Gypsum Plaster over self furring Metal Lath wrapped directly around Column with 3/8" space between Flange and Plaster.
F	2 Hour	4 Each multiple layers of 1/2" Gypsum Wallboard glued to Column Flanges and successive layers. Wallboard layer below outer layer secured to Column with doubled Number 18 Gauge Ties spaced 15" on center.
G	2 Hour	2 Layers of 5/8" Type "X" Gypsum Wallboard screwed to 1-5/8" x 1" x 25 Gauge Channels spaced 16" on center and secured to 1-1/2" Channel Furring. Installed parallel to and on each side of Beam Flanges.

MARK A
Nominal 4-Hour Rating
- 7/8" Portland cement plaster
- Metal lath
- 3/4" Vertical channels

MARK B
Nominal 2-Hour Rating
- 7/8" Portland cement plaster
- 3/4" Air space
- 2½" Total thickness

Table 7-7
Fireproofing Using Spray Applied Albi-Clad™ or Duraspray™

Columns

TYPE OF SYSTEM: Intumescent Mastic
MATERIAL: Albi Manufacturing Division ALBI-CLAD 800. Suitable for exterior use.

Description	Material Cost per 100 sq. ft.	Manhours Each	Labor Cost @ 26.70/hr.	Overhead @ 60% of Labor Cost	Total Cost	Total Including 10% Profit	Subcontract Std. Unit Price per sq. ft.
Nominal 1 Hour Rating - 1/4" Thick							
Type 1 Structural Steel	$700.00	(5.00)	$133.50	$80.10	$913.60	$1,004.96	$10.05
Type 2 Structural Steel	700.00	(5.30)	141.51	84.91	926.42	1,019.06	10.19
Type 3 Structural Steel	700.00	(5.80)	154.86	92.92	947.78	1,042.56	10.43
Nominal 2 Hour Rating - 1/2" Thick							
Type 1 Structural Steel	1,400.00	(10.00)	267.00	160.20	1,827.20	2,009.92	20.10
Type 2 Structural Steel	1.400.00	(10.60)	283.02	169.81	1,852.83	2,038.11	20.38
Type 3 Structural Steel	1,400.00	(11.60)	309.72	185.83	1,895.55	2,085.11	20.85

TYPE OF SYSTEM: Intumescent Mastic
MATERIAL: Albi Manufacturing Division ALBI-CLAD. For interior use only.

Description	Material Cost per 100 sq. ft.	Manhours Each	Labor Cost @ 26.70/hr.	Overhead @ 60% of Labor Cost	Total Cost	Total Including 10% Profit	Subcontract Std. Unit Price per sq. ft.
Nominal 1 Hour Rating - 1/6" Thick							
Type 1 Structural Steel	$700.00	(5.00)	$133.50	$80.10	$913.60	$1,004.96	$10.05
Type 2 Structural Steel	700.00	(5.30)	141.51	84.91	926.42	1,019.06	10.19
Type 3 Structural Steel	700.00	(5.80)	154.86	92.92	947.78	1,042.56	10.43
Nominal 2 Hour Rating - 1/8" Thick							
Type 1 Structural Steel	1,400.00	(10.00)	267.00	160.20	1,827.20	2,009.92	20.10
Type 2 Structural Steel	1,400.00	(10.60)	283.02	169.81	1,852.83	2,038.11	20.38
Type 3 Structural Steel	1.400.00	(11.60)	309.72	185.83	1,895.55	2,085.11	20.85

Definitions:
Type 1 Structural Steel has up to 2 Square Feet Per Lineal Foot
Type 2 Structural Steel has over 2 Square Feet to 5 Square Feet Per Lineal Foot
Type 3 Structural Steel has over 5 Square Feet Per Lineal Foot
Notes:
1. Steel must be primer coated to receive Fireproof Coatings above. Cost of primer coat is not included.
2. For Integral Color Top Coat, add $1.00 Per Square Foot.
3. Subcontract Standard Unit Prices include required fibrous glass reinforcement at flange edges on nominal Two Hour Rated Systems.
4. Subcontract Standard Unit Prices are based on a minimum installation of 5,000 Square Feet.
5. For less than 5.000 Square Feet, add 15% to Standard Subcontract Unit Prices.
6. Use an appropriate formula for calculating Square Foot Quantities.
7. The Subcontract Standard Unit Prices are based on a maximum floor to top of column height of 16'0".
8. For work performed above 16'0", add 10% for each additional 4'0" of height or fraction thereof.
9. ALBI-CLAD 800" and ALBI-CLAD" are manufactured by ALBI Manufacturing Division of StanChem, (860)828-0571.
Other Types of Fireproofing:
1. Poured Concrete Fireproofing of Structural Steel. Poured Concrete Fireproofing of Structural Steel may be estimated using data in Concrete Accounts. See Account 3-0.
2. Gunite (Air Blown Concrete) Spray-On Fireproofing of Steel Plate.
(a) Non-reinforced: $1.20 Per Square Foot Per Inch of Thickness.
(b) Reinforced with 2" x 2" Mesh attached to Studs or Tied to Nuts: $1.80 Per Square Foot Per Inch of Thickness of Gunite.

Source: Smith, Materials of Construction, McGraw-Hill Book Co.

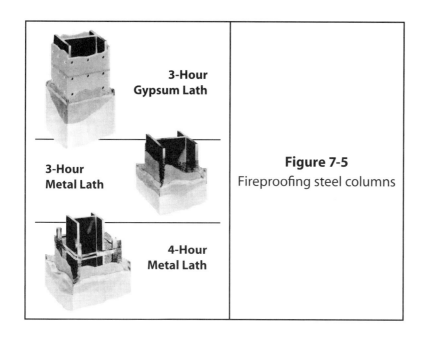

Figure 7-5
Fireproofing steel columns

Table 7-8
Coefficients of Heat Transmission

Material	Conductivity k	Conductance C	Resistance R(1/C)
Air, inside film:			
Heat flow up		1.64	0.61
Heat flow horizontal		1.46	0.68
Heat flow down		1.08	0.92
Air, outside film:			
7.5 mi/h (12.1 km/h)		4.0	0.25
15 mi/h (24.1 km/h)		6.0	0.17
Air, enclosed, in (19 mm) or more vertical		1.08	0.92
Insulation:			
Blankets, batts, or loose fill, mineral or vegetable	024		4.16
Rigid insulation board, wood or vegetable fiber	0.33		3.03
Vermiculite or Perlite	0.48		2.08
Roofing:			
Asbestos shingles		4.75	0.21
Asphalt shingles		2.27	0.45
Wood shingles		1.06	0.95
Built-up roofing		3.00	0.33
Slate, 1/2 in (12.7 mm)		20.00	0.05
Exterior materials:			
Stucco, 1 in (25 mm)	5.00	5.00	0.20
Wood siding, 34 in (19 mm)	0.80	1.06	0.93
Face brick, 4 in (102 mm)	9.00	2.25	0.44
Common brick, 4 in (102 mm)	5.00	1.25	0.80
Douglas fir, Ta in (19 mm)	0,80	1.06	0.93
Southern pine, 11 in (38 mm)	0.80	0.53	1.88
2 in (51 mm)	0.80	0.40	2.50
3 in (76 mm)	0.80	0.27	3.70
Exterior materials:			
Concrete, sand, and gravel	12.00		0.08
Vermiculite or Concrete, 1 in (25 mm)	0.86	0.86	1.16
Stone	12.50		
Glass, 1/4 in (6.3 mm)	6.00		0.043
1/8 in (3.2 mm)	6.00		0.021
Interior materials:			
Gypsum lath and plaster		2.44	0.41
Metal lath and ¾-in (19-mm) plaster		6.66	0.15
Vermiculite or Perlite plaster 1 in (25 mm)	1.7		0.59
Plywood, 3/8 in (9.5 mm)		2.10	0.46
In the metric system, U value = Btu (h)(f12)(°F)			
(= 5.678 W/(m2)(°C)			

Source: Watson, Construction Materials and Processes, 2d. ed., © McGraw-Hill Book Co.

WATER-RESISTANCE

One of the main purposes for plastering a building is to protect the interior from the environment (heat, cold, and moisture). Proper design, the selection and use of appropriate materials, and professional construction practices are all required to produce an effective weather-resistant design.

Unlike gypsum plaster or gypsum board (which are used on interior surfaces), Portland cement plaster will not dissolve in the presence of water. For that reason, it can be used on the exterior of buildings, but with special care.

- Wind can drive water through a stucco finish, by the force of air pressure differences.
- An exterior stucco finish derives its ability to protect the interior from water intrusion not only from the substantial material of the cement plaster coats, but also from the presence of and materials used in the weather-resistant barrier and underlying sheathing materials.

To prevent water intrusion in a stucco-finished building, a weather-resistant barrier must be provided under the cement plaster.

- As we've seen, this barrier is typically provided by Grade D Kraft building paper, asphalt-saturated building felt or polyolefin fabric. Over wood-based sheathing, two layers of Grade D paper or polyolefin fabric–or one layer of Grade D paper and one layer of polyolefin fabric–are required.
- The separate sheets of paper used in the weather-resistant barrier must be properly overlapped (shingled) top to bottom and side to side.
- Any water that manages to penetrate the cement plaster layers is stopped by the weather-resistant barrier, and flows down and away from the building.
- To complete the weather-resistant barrier, proper flashing and accessories must be provided around all edges–such as weep screeds along the bottom of all wall surfaces, and flashing around all windows and doors.

Portland cement plaster is sometimes used on interior surfaces, especially when there is regular exposure to water (e.g., showers, bathtubs and rooms of high humidity). In those cases, the Portland cement plaster does not typically serve as the finish material, but simply as the base to a truly waterproof material such as ceramic tile.

> **DEFINITION**
>
> *"Waterproofing"* is the process of providing a surface with near-total water repellency, preventing the free flow of liquid water. Stucco is not waterproof.
>
> *"Damp-proofing"* is the process of providing a surface with a measure of controlling the capillary passage of water, water vapor or moisture through a structural surface. An example of damp-proofing would be the use of a hot or cold mopping of bituminous coating applied to concrete or masonry basement walls prior to plastering.

ACOUSTIC PROPERTIES

Buildings located near highways, railroads, and airports, and in densely populated urban areas can be exposed to unwanted noise. To make such structures acceptable as residences of places of business, some degree of sound insulation should be provided.

The "loudness" of sound depends on two things: (1) the frequency or pitch, and (2) the intensity.

- Loudness is measured in units called "decibels" (dB).
- An intensity of 0 dB is the least sound discernible to the humane ear.
- An intensity of 100 dB is equal to the sound of close-by thunder or a passing train, and is considered deafening (allows no other sound to be heard).

In construction terms, we are concerned with controlling the sound that reverberates within a room and the sound that is transmitted into or out of adjacent rooms.

- Controlling the sound within a room depends on the sound absorptiveness and reflectivity of a material.
- The transmission of sound into or from adjacent rooms depends on the transmissibility of the material or assembly of materials.

The acoustic absorptiveness and reflectivity of a surface material are rated and reported as the material's "noise reduction coefficient" (NRC).

- The NRC of a surface ranges from zero (a surface that absorbs no sound) to 1 (a surface that absorbs all sounds striking it).
- An NRC rating of 0.50 indicates a surface that reflects 50% of the sound striking it and absorbs 50%.

The transmission of sound through a wall or floor is controlled by proper design of the structure and the use of sound-deadening materials and assemblies. The "sound transmission class" (STC) of a panel is determined according to ASTM E 90.

- An STC rating of 25 to 30 dB is poor; normal speech is distinctly heard and understood through the wall.
- An STC of 50 dB or more is excellent; very loud noises are faintly heard or are inaudible.

Acoustically absorptive materials consist of:

ACOUSTIC CEILING TILES

- "Tiles" are typically 12" x 12" (0.3 X 0.3 m). "Panels" are 24" x 36" or larger.
- Made of sugarcane, wood, mineral, or glass fiber felt, and compressed into boards of various thicknesses.
- They may be perforated with round or slotted holes, or otherwise textured to improve their absorptive properties.

SPRAYED ACOUSTIC PLASTER

- Popularly called "cottage cheese" or "popcorn" ceilings.
- Made of perlite or vermiculite aggregate with gypsum plaster.
- Bagged, premixed acoustical plasters are available.

SPRAYED FIBERS

- Mineral wool fibers with a binder packed in bags.
- An adhesive is first applied to the surface to be treated; then the fibers are sprayed onto the primed surface.

THE ACOUSTICS OF TYPICAL WALLS

Wall assemblies using different framing configurations and different *interior* wall materials are shown below, with their corresponding decibel ratings. Studied wall configurations are not necessarily stucco-finished on the exterior, although they can be (with either insulating board or ¾-inch sheathing). These figures are provided here, then, as a general guide.

	Wall Assembly Materials	dB
1	Wood stud framing with ¾-inch insulating board on each side	32 dB
2	Wood stud framing with ¾-inch insulating board and ½-inch plaster on each side (not recommended)	45 dB
3	Wood stud framing with 3/8 inch gypsum lath and ½-inch plaster on each side	34 dB
4	Two separate walls of 2 x 2 wood studs with ½-inch insulating board between and on each side (not recommended)	43 dB
5	Staggered 2 x 4 studs with ½-inch insulating board and ½-inch plaster on each side	50 dB

CHAPTER 8

Cost Estimating

IMPORTANT NOTE: Cost figures are provided here as examples only. Updated information for your local area/region should be developed before you attempt to create a cost estimate for stucco work.

ESTIMATING STUCCO WORK

Cost estimating is an important part of a successful stucco job. The stucco estimate is a combination of *material costs* with *labor costs*, to get *direct material and labor costs*. *Indirect costs* should then be added to cover *overhead items* required to do the work. Finally, *profit* is determined and added to the total direct and indirect costs.

Construction estimating is not an absolute science; instead, it is more like the thoughtful prediction of costs. Predicting the exact quantities and costs of materials that will be installed by workers, and predicting exactly how long that installation will take, is a complex task. For example, installed material quantities can vary from the estimated quantities due to any of the following factors:

- Design changes.
- Unforeseen site problems (obstacles, clutter, access problems, etc.)
- Delays caused by weather, delayed material deliveries, other contractors.
- Price changes.
- Waste.

Actual man-hours can vary from estimated man-hours, depending on the experience of the actual crew involved in a project.

On the following pages, we'll provide some aids to creating an estimate, including hourly labor costs and scaffolding rental rates. In addition to the tables we offer here, though, you should also include equipment rental costs, power/fuel costs, tools (a percentage of which will be lost or broken on a typical job), etc.

Remember: no two projects are ever exactly the same, and no estimate every matches the final result. The best approach, then, when developing a profitable estimate is to build a reasonable margin for such contingencies into the estimate.

WHO WILL DO WHAT?

Table 8-1
Scope of Work Responsibilities

This table clarifies the responsibilities of the parties involved in the construction project with a stucco cladding. The parties referred to are the architect/owner, the general contractor/builder and the wall and ceiling/plastering contractor.

Scope of Work	Architect/Owner	General Contractor	W/C
1. Construction documents design, drawings and specifications	x		
2. Site conditions for storage and setting up of plastering operation		x	
3. Qualified plastering contractor (experience in scope of work and trained labor force)			x
4. Awarding the work to a qualified plastering contractor (experience in scope of work and trained labor force)		x	
5. Pre-job meeting to review all aspects of the stucco system and interface with other building components	x	x	x
6. Scheduling and coordination of all trades		x	
7. Tenting and heating for cold weather conditions*		x	
8. Buildings, substrate, framing, sheathing, concrete, CMU*		x	
9. Finish samples (color and texture)			x
10. Approval of finish samples (color and texture)	x		
11. Submittals			x
12. Approval of submittals	x		
13. Metal flashing, caps		x	
14. Windows/doors — flashing and installation		x	
15. Roof, balcony — flashing, caps, etc.		x	
16. Flashing at heads of windows and doors*		x	x
17. Sealants and/or backing rod — at windows, doors or wall penetrations where construction documents indicate*		x	x
18. Coordination of windows, doors and other devices in wall surface*		x	
19. Wall and ceiling penetration units, vents, pipes, railings, electrical items, HVS, etc.		x	
20. Scaffolding and/or other equipment necessary for stucco work*			x
21. Weather-resistant barrier as part of stucco assembly			x
22. Location and type of trim accessory joints (control joints, reveals, expansion joints and soffit vent screeds	x		
23. Notification of unsatisfactory conditions before stucco work			x
24. Correcting unsatisfactory conditions before stucco work	x	x	
25. Lath installation			x
26. Trim accessories: control joints, reveals, expansion joints, casing beads, vent screeds, etc.			x
27. Preparation of CMU and/or concrete surfaces		x	
28. Bonding agent over CMU and/or concrete surfaces			x
29. Wood nailing strips		x	
30. Finished wood trim		x	
31. EPS implants			x
32. All equipment, tools, materials, and labor for stucco system			x
33. Delivery and /or stocking of materials			x
34. Power and water		x	
35. Safe grade conditions		x	
36. Covering for protection from rain for finish work			x
37. Protection from water runoff from roof, etc.		x	
38. Covering and protection of building components from stucco system operations			x
39. All phases of plastering work			x
40. Finish coat			x
41. Paint		x	
42. Sealants		x	x
43. Quality control checklist	x	x	x
44. Clean up — only stucco scope of work			x
45. Protection of finished wall surface from damage after stucco work is completed		x	
46. Protection of stucco phases from other construction operations		x	

*The contractual agreement between the builder/general contractor and the wall and ceiling contractor/plastering contractor will supersede this Scope of Responsibilities table. Therefore, these items must be indicated in the construction documents and reviewed at the pre-job meeting.

SCAFFOLDING - RENTAL RATES

Table 8-2 / Steel Scaffolding and Accessories	
Scaffolding	Rental Rates per Month
Frames, All sizes including braces, pins, base plates, guard rail posts & rails	$22.00 per Frame
Hoist Arms	13.00 each
"T" Hoist & Gooseneck	18.00 each
Planking, Add to basic scaffold cost shown above	
Wood 2" x 12"	0.31 per sq. ft.
Aluminum scaffold plank	0.32 per sq. ft.
Steel Truss scaffold plank	0.35 per sq. ft.
Casters. Steel and Rubber	
5", 6", 8"	10.00 each
12"	10.00 each
Side Brackets	2.50 each
Adjustable Screw Jacks 2.00 each	
Outriggers	7.00 each
Stair Units 5' or 6' with guard rails	25.75 each
Access Ladder 6'	3.25 each
Putlogs: Parallel Truss 8' or 12'	12.00 each
Perimeter Guard Rail Post	6.00 each

Notes:

For scaffolding Frames rented in large quantities and for long periods, rates may be reduced as much as 50%. To estimate the rental cost of Tubular Steel Scaffolding the following information must be known:

1. (a) The length of the scaffolding; (b) The width of the scaffolding; (c) The height of the scaffolding.

   With this information known, the following formula may be used to determine the quantity of "frames" that will be required.

$$\left(\frac{\text{Scaffold Length in feet}}{10} + 1 \right) \times \left(\frac{\text{Scaffold Width in feet}}{4} \right) \times \left(\frac{\text{Scaffold Height in feet}}{6.5} \right) = \text{Quantity of Frames}$$

2. The square footage of the working platform(s) on the scaffold (planked area) must also be known. This is usually estimated as width of platform required x length of the scaffold x quantity of platforms required on the scaffold.

3. It should be noted that various sizes of scaffolding may be required on a given project. These various sizes and quantities of each must be determined from the job being estimated.

4. The length of time the scaffolding is to be used must also be known. For estimating purposes use one month as minimum rental period and calculate all other time in increments of one month. (6 weeks would equal 2 months rental.)

Example:

> 1. Assume scaffolding required is (a) 20 feet long; (b) 4 feet wide;
>   (c) 13 feet high: the total frames would be estimated as...
>
> $\left(\frac{20}{10}+1\right) \times \left(\frac{4}{4}\right) \times \left(\frac{13}{6.5}\right) = 3 \times 1 \times 2 = 6 \text{ frames}$
>
> 2. Assume one (1) working platform 4 feet wide is required on one (1) level of the scaffold; the square footage of planked area would be estimated as 4′ x 20′ = 80 Square Feet.
> 3. For this example, assume only one (1) scaffold of the size shown above is required.
> 4. Assume scaffold usage will be six weeks (use 2 months as rental period per instructions).

Rental Costs, for the example:

Frames braces, pins, base plates, guard rails & posts, 6 EA @ $22.00	$132.00	per month
Planking, steel truss 80 SQ. FT. @ $0.35	28.00	per month
Estimated Rental Cost, Per Month	$160.00	per month
Estimated Rental Period	x 2	months
Total Estimated Costs	$320.00	

ROLLING SCAFFOLD

For estimating purposes the width of a rolling scaffold should be calculated as the scaffold height divided by 4; the length of a rolling scaffold is usually limited to 20′ maximum. Using this formula as a basis for determining the quantity of "frames" required and for assigning square footage for platform(s), rolling scaffolds can be estimated using the same procedure as shown plus the additives for casters (2 per frame) and access ladders as may be required.

HEAVY-DUTY ADJUSTABLE SHORING SYSTEM

Adjust-a-shore Towers of any height can be assembled using base frames with extension frames at top or bottom to provide adjustments at 1′ intervals. Capacity 11,000 LBS Per Leg; 44,000 LBS Per Tower.

Table 8-3 / Shores and Accessories	
Description	Rental Per Month
Frames, any size base frames and extension frames including braces, pins and clamps	$22.00
Screwjacks, including base plates (fixed or swivel) and "U" Heads	9.00

SHORES

Adjustable Vertical Post Shores with heights from 6 to 15 Feet using one 6 Foot High base section with three interchangeable top sections, each having and adjustment range of about 5 Feet.

Table 8-4 / Scaffolding Rental

Exterior scaffolding, tubular steel, 60" wide, with 6'4" open end frames set at 7', with cross braces, base plates, mud sills, adjustable legs, post-mounted guardrail, climbing ladders and landings, brackets, clamps and building ties. Including two 2" x 10" scaffold planks on all side brackets and six 2" x 10" scaffold planks on scaffold where indicated. Add local delivery and pickup. Minimum rental is one month.

Costs are per square foot of building wall covered, per month.

	Craft @ Hrs.	Unit	Matl.	Labor	Total
With plank on scaffold only	--	SF	.65	--	.65
With plank on side brackets and scaffold	--	SF	.90	--	.90

HOURLY LABOR COSTS

Table 8-5 / Plastering, Subcontract

See also Lathing. Typical costs including material, labor, equipment and subcontractor's overhead and profit.

	Craft @ Hrs.	Unit	Matl.	Labor	Total
Acoustical plaster					
Including brown coat, 1-1/4" total thickness	--	SY	--	--	19.00
Plaster on concrete or masonry, with bonding agent					
Gypsum plaster, 1/2" sanded	--	SY	--	--	10.00
Plaster on interior wood frame walls, no lath included					
Gypsum plaster, 7/8", 3 coats	--	SY	--	--	15.00
Keene's cement finish, 3 coats	--	SY	--	--	16.50
Cement plaster finish, 3 coats	--	SY	--	--	15.50
Gypsum plaster, 2 coats, tract work	--	SY	--	--	13.00
Add for ceiling work	--	%	--	--	15.0
Plaster for tile					
Scratch coat	--	SY	--	--	7.50
Brown coat and scratch coat	--	SY	--	--	11.50
Exterior plaster (stucco)					
Walls, 3 coats, 7/8", no lath included					
Textured finish	--	SY	--	--	11.00
Float finish	--	SY	--	--	13.00
Dash finish	--	SY	--	--	14.50
Add for soffits	--	%	--	--	15.0
Portland cement plaster, including metal lath					
Exterior, 3 coats, 7/8" on wood frame					
Walls	--	SY	--	--	18.00
Soffits	--	SY	--	--	19.00
Deduct for paper back lath	--	SY	--	--	-.16

Table 8-6 / Lathing Estimate

	Craft @ Hrs.	Unit	Material	Labor	Total
Lath					
Gypsum lath, nailed to walls and ceilings					
3/8" x 16" x 48"	BR@.076	SY	2.60	1.86	4.46
1/2" x 16" x 48"	BR@.083	SY	2.85	2.03	4.88
Add for foil back insulating lath	--	SY	.62	--	.62
Steel lath, diamond pattern (junior mesh), 27" x 96" sheets, nailed to walls					
2.5 lb galvanized	BR@.076	SY	2.07	1.86	3.93
3.4 lb galvanized	BR@.076	SY	2.38	1.86	4.24
Wood lath. See Lumber, Redwood					
Z-Riblath, 1/8" to wood frame (flat rib)					
2.75 lb painted, 24", 27" x 96" sheets	BR@.076	SY	3.11	1.86	4.97
3.4 lb painted, 24", 27" x 96" sheets	BR@.088	SY	3.78	2.15	5.93
Riblath, 3/8" to wood frame (high rib)					
3.4 lb painted, 24", 27" x 96" sheets	BR@.088	SY	2.54	2.15	4.69
3.4 lb galvanized, 24", 27" x 96" sheets	BR@.088	SY	3.00	2.15	5.15
3.4 lb painted paperback, 27" x 96" sheets	BR@.088	SY	2.95	2.15	5.10
Add for ceiling applications	--	%	--	25.0	--
Corner bead, 26 gauge, 8' to 12' lengths. Use 100 LF as a minimum job charge					
Flexible all-purpose bead	BR@.027	LF	.46	.66	1.12
Expansion, 2-1/2" flange	BR@.027	LF	.48	.66	1.14
Cornerite or cornerlath					
2" x 2" x 4", steel-painted copper alloy	BR@.027	LF	.16	.66	.82
Corneraid					
2-1/2" smooth wire	BR@.027	LF	.26	.66	.92
Casing beads					
Square nose, short flange, 1/2" or 3/4"	BR@.025	LF	.36	.61	.97
Quarter round, short flange, 1/2" or 3/4"	BR@.025	LF	.34	.61	.95
Square nose, expansion, 1/2" x 3/4"	BR@.025	LF	.41	.61	1.02
Quarter round, expansion, 1/2" x 3/4"	BR@.025	LF	.41	.61	1.02
Archbead (flexible plastic nose) for curves	BR@.025	LF	.57	.61	1.18
Base screed, flush or curved point	BR@.018	LF	.41	.44	.85
Expansion joint, 26 gauge, 1/2" ground	BR@.025	LF	1.04	.61	1.65
Archaid (for curves and arches), 8' length	BR@.028	LF	.47	.68	1.15
Galvanized stucco netting					
1" x 20 gauge x 48", 450 SF/roll	--	Roll	98.30	--	98.30
1" x 18 gauge x 48", 450 SF/roll	--	Roll	152.00	--	152.00
1" x 18 gauge x 48", 600 SF/roll	--	Roll	191.00	--	191.00
1-1/2" x 17 gauge x 36", 150 SF/roll	--	Roll	54.30	--	54.30
Self-furring	--	Roll	54.30	--	54.30
Self-furring, paperback, 100 SF/roll	--	Roll	60.00	--	60.00
LATHING, SUBCONTRACT					
Metal lath, nailed on wood studs					
2.5 lb diamond mesh, copper alloy	--	SY	--	--	4.86
2.5 lb galvanized	--	SY	--	--	5.43
3.4 lb painted copper alloy	--	SY	--	--	5.80
Add for ceiling applications	--	%	--	--	15.0
Gypsum lath, nailed on wood studs					
3/8" thick	--	SY	--	--	5.59
1/2" thick	--	SY	--	--	6.00
Add for ceiling applications	--	%	--	--	15.0
Add for foil facing	--	SY	--	--	.93
15 lb. felt and stucco netting	--	SY	--	--	3.11

Table 8-7 / Craft Direct Cost

Craft	Base wage per hour	Taxable fringe benefits (@5.15% of base wage)	Insurance and employer taxes (%)	Insurance and employer taxes ($)	Non-taxable fringe benefits (@4.55% of base wage)	Total hourly cost used in this book
Bricklayer	$19.75	$0.92	31.35%	$5.92	$0.82	$25.61
Bricklayer's Helper	13.77	0.71	31.35%	4.54	0.63	19.65
Building Laborer	13.94	0.72	33.05%	4.85	0.63	20.14
Carpenter	16.93	0.87	31.95%	5.69	0.77	24.26
Cement Mason	17.28	0.89	26.86%	4.88	0.79	23.84
Drywall installer	17.56	0.90	28.32%	5.23	0.80	24.49
Drywall Taper	17.56	0.90	28.32%	5.23	0.80	24.49
Electrician	19.86	1.02	24.45%	5.10	0.90	26.88
Floor Layer	16.74	0.86	32.55%	5.73	0.76	24.09
Glazier	16.11	0.83	29.52%	5.00	0.73	22.67
Lather	17.95	0.92	25.05%	4.73	0.82	24.42
Marble Setter	16.07	0.83	23.35%	3.95	0.73	21.58
Millwright	17.48	0.90	31.95%	5.87	0.80	25.05
Mosaic & Terrazzo Worker	17.05	0.88	23.35%	4.19	0.78	22.90
Operating Engineer	20.11	1.04	26.10%	5.52	0.91	27.58
Painter	17.92	0.92	35.25%	6.64	0.82	26.30
Plasterer	17.84	0.92	32.35%	6.07	0.81	25.64
Plasterer Helper	13.62	0.70	32.35%	4.63	0.62	19.57

Table 8-8 / Typical General Contractor's Indirect Costs

COST PERCENTAGES FOR ONE STORY/20' HEIGHT CONSTRUCTION WORK

Main Account	Construction Equipment Including FOLM	Field Supervision	Temporary Construction	Small Tools & Consumeables	FICA, Unemployment, Workers Comp. Insurance	Contractor Home Office Costs	Total Indirect Costs
2 - 0*	8%	5%	4%	5%	17%	11%	50%
3 - 0	9%	6%	4%	5%	22%	13%	59%
4 - 0	6%	4%	3%	4%	22%	13%	52%
5 - 0	28%	7%	7%.	5%	28%	15%	90%
6 - 0	2%	3%	2%	3%	22%	11%	43%
7 - 0	3%	3%	3%	3%	22%	11%	45%
8 - 0	2%	3%	2%	3%	22%	11%	43%
9 - 0	9%	5%	4%	5%	24%	13%	60%
10 - 0	3%	3%	2%	3%	19%	11%	41%
15 - 0	18%	5%	5%	5%	26%	14%	73%
16 - 0	17%	5%	4%	55	25%	14%	70%
100 - 0**	30%	5%	4%	5%	28%	13%	85%

TYPICAL GENERAL CONTRACTOR INDIRECT COST PERCENTAGES PER STORY FOR CONSTRUCTION WORK

Main Account	1 Story 20'0"	2 Stories 36'0"	3 Stories 52'0"	4 Stories 68'0"	5 Stories 84'0"	6 Stories 100'0"
2 - 0*	50%	50%	50%	50%	50%	50%
3 - 0	59%	62%	65%	68%	71%	74%
4 - 0	52%	55%	58%	61%	64%	67%
5 - 0	90%	97%	104%	121%	118%	125%
6 - 0	43%	46%	49%	52%	55%	58%
7 - 0	45%	48%	51%	54%	57%	60%
8 - 0	43%	46%	49%	52%	55%	58%
9 - 0	60%	69%	78%	87%	96%	105%
10 - 0	41%	47%	53%	59%	65%	71%
15 - 0	73%	75%	77%	79%	81%	83%
16 - 0	70%	72%	74%	76%	78%	80%
100 - 0**	85%	90%	95%	100%	105%	110%

* Overall project height generally has no effect on sitework costs; thus, Indirect percentages for Account 2 - 0 remain at 50%.
** Main Account 100 - 0 is found only in Volume 4 of the Process Plant Construction Estimating Standards.

Table 8-9 / Dollars per Manhour Allowance for Indirect Job Costs

EXCLUDING GENERAL CONTRACTOR PROFIT, FOR BUILDINGS AND/OR PROJECTS WITH MAXIMUM STORIES OR HEIGHT NOT EXCEEDING:

Main Account Number	Average of Wage Rate	% of Wage Rate	One Story 20'0"	% of Wage Rate	Two Stories 36'0"	% of Wage Rate	Three Stories 52'0"	% of Wage Rate	Four Stories 68'0"	% of Wage Rate	Five Stories 84'0"	% of Wage Rate	Six Stories 100'0"
2 - 0*	N/A	N/A	N/A	N/A	N/A	N/A	N/A	N/A	N/A	N/A	N/A	N/A	N/A
3 - 0	$26.40	59%	$15.58	62%	$16.37	65%	$17.16	68%	$17.95	71%	$18.74	74%	$19.54
4 - 0	25.65	52%	13.34	55%	14.11	58%	14.88	61%	15.65	64%	16.42	67%	17.19
5 - 0	29.70	90%	26.73	97%	28.81	104%	30.89	111%	32.97	118%	35.05	125%	37.13
6 - 0	27.45	43%	11.80	46%	12.63	49%	13.45	52%	14.27	55%	15.10	58%	15.92
7 - 0	25.35	45%	11.41	48%	12.17	57%	12.93	54%	13.69	57%	14.45	60%	15.21
8 - 0	27.45	43%	11.80	46%	12.63	49%	13.45	52%	14.27	55%	15.10	58%	15.92
9 - 0	26.60	60%	15.96	69%	18.35	78%	20.75	87%	23.14	96%	25.54	105%	27.93
10 - 0	27.45	41%	11.25	47%	12.90	53%	14.55	59%	16.20	65%	17.84	71%	19.49
15 - 0	29.40	73%	21.46	75%	22.05	77%	22.64	79%	23.23	81%	23.81	83%	24.40
16 - 0	31.30	70%	21.91	72%	22.54	74%	23.16	76%	23.79	78%	24.41	80%	25.04
100 - 0**	29.55	85%	25.12	90%	26.60	95%	28.07	100%	29.55	105%	31.03	110%	32.51

* Overall project height generally has no effect on sitework costs; thus, Indirect percentages for Account 2 - 0 remain at 50%.
** Main Account 100 - 0 is found only in Volume 4 of the Process Plant Construction Estimating Standards.

Table 8-10 / Regional Modifiers

State	Modifier	State	Modifier	State	Modifier
Alaska	1.42	Kentucky	0.95	Ohio	1.00
Alabama	0.86	Louisiana	0.89	Oklahoma	0.85
Arizona	0.99	Maine	1.00	Oregon	1.07
Arkansas	0.86	Maryland	0.95	Pennsylvania	
California		Massachusetts	1.08	Philadelphia	1.19
Los Angeles	1.15	Michigan	1.01	Other	1.03
San Francisco	1.29	Minnesota	1.06	Rhode Island	1.07
Other	1.05	Mississippi	0.87	South Carolina	0.89
Colorado	1.10	Missouri	0.94	South Dakota	0.97
Connecticut	1.06	Montana	0.98	Tennessee	0.91
Delaware	1.06	Nebraska	0.92	Texas	0.87
District of Columbia	1.02	Nevada	1.08	Utah	0.97
Florida	0.90	New Hampshire	0.94	Vermont	1.03
Georgia	0.89	New Jersey	1.12	Virginia	0.89
Hawaii	1.46	New Mexico	0.90	Washington	1.11
Idaho	1.01	New York		West Virginia	1.05
Illinois	1.06	New York City	1.38	Wisconsin	1.03
Indiana	1.00	Other	1.04	Wyoming	0.99
Iowa	1.03	North Carolina	0.88		
Kansas	0.91	North Dakota	1.02		

Table 8-11 / Concrete Pumping & Temporary Enclosures Rentals

	Craft@Hrs	Unit	Material	Labor	Total
Concrete Pumping with a Boom Truck, Subcontract					
Includes truck rent, operator, local travel but no concrete. Add costs equal to 1 hour for equipment setup and 1 hour for cleanup. Use 4 hours as the minimum cost for 23, 28 and 32 meter boom trucks and 5 hours as the minimum cost for 36 through 52 meter boom trucks. Estimate the actual pour rate at 70% of the rated capacity on thicker slabs and 50% of the capacity on most other work.					
Boom lengths over 23 meters include both an operator and an oiler. Costs shown include subcontractor's markup.					
23 meter boom (75'), 70 CY per hour rating	--	Hr	--	--	55.00
Add per CY pumped with 23 meter boom	--	CY	--	--	2.00
28 meter boom (92'), 70 CY per hour rating	--	Hr	--	--	65.00
Add per CY pumped with 28 meter boom	--	CY	--	--	2.00
32 meter boom (105'), 90 CY per hour rating	--	Hr	--	--	75.00
Add per CY pumped with 32 meter boom	--	CY	--	--	2.00
36 meter boom (120'), 90 CY per hour rating	--	Hr	--	--	95.00
Add per CY pumped with 36 meter boom	--	CY	--	--	2.00
42 meter boom (138'), 100 CY per hour rating	--	Hr	--	--	125.00
Add per CY pumped with 42 meter boom	--	CY	--	--	2.50
52 meter boom (170'), 100 CY per hour rating	--	Hr	--	--	150.00
Add per CY pumped with 52 meter boom	--	CY	--	--	3.00
Temporary Enclosures					
Rented chain link fence and accessories. Costs are a one-time charge for up to six months usage on a rental basis. Costs include installation and one trip for removal and assume level site with truck access along pre-marked fence line. Add for gates and barbed wire as shown. Minimum charge is $295.					
Chain link fence, 6' high					
Less than 250 feet	--	LF	--	--	1.00
250 to 500 feet	--	LF	--	--	.90
501 to 750 feet	--	LF	--	--	.80
751 to 1,000 feet	--	LF	--	--	.70
Over 1,000 feet	--	LF	--	--	.65
Add for gates					
6' x 10' single	--	Ea	--	--	77.50
6' x 12' single	--	Ea	--	--	100.00
6' x 15' single	--	Ea	--	--	130.00
6' x 20' double	--	Ea	--	--	150.00
6' x 24' double	--	Ea	--	--	200.00
6' x 30' double	--	Ea	--	--	250.00
Add for barbed wire (per strand, per lineal foot)	--	LF	--	--	.15
Contractor furnished items, installed and removed, based on single use and no salvage value					
Railing on stairway, two sides, 2" x 4"	CL@.121	LF	.78	3.54	4.32
Guardrail at second and higher floors: toe rail, mid rail and top rail	CL@.124	LF	2.08	3.62	5.70
Plywood barricade fence					
Bolted to pavement, 8' high	CL@.430	LF	7.79	12.60	20.39
Post in ground, 8' high	CL@.202	LF	8.68	5.90	14.58
8' high with 4' wide sidewalk cover	CL@.643	LF	22.20	18.80	41.00

Source for Tables 18-1 through 18-10: Richardson Engineering Services, "Process Plant Estimating Standards," 2002.

CHAPTER 9

Stucco Maintenance and Repair

IDENTIFYING DAMAGE

Even the most carefully mixed and most expertly applied Portland cement plaster will develop cracks. While the proper use of control and expansion joints, addition of fiber shorts to basecoat mixes, and using only the best quality ingredients (especially sand) can go a long way toward limiting the amount and severity of cracking that will occur–nevertheless, stucco will crack as it dries, and over time, as it contracts and expands. The real issue is whether any cracking is excessive, and whether any other deficiencies develop in a stucco finish. So this chapter is concerned not so much with "normal" cracking, but with actual problems in a stucco finish.

Before beginning any repair, the stucco should be thoroughly inspected to evaluate the extent of damage and determine how much material must be repaired or replaced. Some damage will be obvious–for example, holes, missing sections or layers of plaster. Other damage may be less obvious in a visual inspection, but can be discovered by lightly tapping on the surface. For example, water-damaged stucco often forms bulges as the coats of cement plaster begin to delaminate and the metal lath and nails rust. The bulge grows as the problem grows; eventually, the damaged section can fall from the building. Such problem areas will make a hollow sound when lightly tapped with a hammer.

- Examination and analysis of samples of the plaster following the procedures outlined by ASTM C 1324 can help you determine its mix components and can provide other relevant information, such as the quality of the bond between coats. This information can offer clues as to the expected useful life of the plaster that remains and aid in determining the appropriate scope of any repair project.

Concrete surfaces can flake or spall for one or more of the following reasons:

- In areas of the country that are subjected to freezing and thawing, the concrete should be air-entrained to resist flaking and scaling of the surface. (Most stucco mixes create an air-entrained product when applied by hand; if you are applying stucco by machine, you may want to use Type A air-entrained cement.) If air-entrained cement is not used, there will be subsequent damage to the surface.
- The water/cement ratio should be as low as possible to improve durability of the surface. Too much water in the mix will produce a weaker, less durable concrete that will contribute to early flaking and spalling of the surface.
- The finishing operations should not begin until the water sheen on the surface is gone and excess bleed water on the surface has had a chance to evaporate. If this excess water is worked into the concrete because the finishing operations are begun too soon, the concrete on the surface will have too high a water content and will be weaker and less durable.
- When construction scaffolding and standoffs are removed/detached at the end of a project, significant areas of plaster, lath and weather barrier can be damaged or even destroyed. This damage is not always visible, yet it can become a major path of water intrusion. Proprietary devices are available that will prevent such damage.

Other causes of stucco failure can include:

- The paper or lath around windows and other openings, or at control or expansion joints was installed incorrectly. The paper must be installed behind accessories at the top and side edges of a wall surface; in front of accessories located along bottom edges. Lath should be wired to the accessory at least every 7".
- Window head flashing is absent or inadequate. Windows with flanges are not "self-flashed"... you need to install flashing.
- One layer of paper was used in the weather-resistant barrier. Most locales require two layers of paper (or an approved housewrap and one layer of paper) over wood-based sheathing, for a weather-resistant barrier.
- Water may be leaking through tears or holes in the paper.
- The windows themselves may be leaking, allowing water to enter the interior and/or behind the weather-resistant barrier.
- Flashing at wall/roof intersections (i.e., where the roofline does not extend below the wall) was not installed at all, or not installed correctly.
- The ledger board for an attached deck or balcony was not properly flashed.
- Moisture was trapped in the wall during construction, due to rain or wet building materials (construction moisture).

- Interior moisture is permeating into the wall as vapor, but the weather-resistant barrier is not permeable enough to allow its escape. In addition, stucco itself creates a very "tight" wall, making the vapor permeability of any housewrap and/or paper/felt all the more important.
- Type 15 felt may be acting as a vapor retarder trapping moisture in the wall. Many believe that using Grade D Kraft for the weather-resistant barrier allows better vapor transmission.
- Oriented strand board (OSB) sheathing has a low perm rating and it may be acting as a vapor retarder, creating a double vapor retarder situation. In addition OSB absorbs and retains moisture making it vulnerable to mold and rot.
- The staples that stick through the sheathing are collecting frost or condensation and dripping within the wall cavity.
- The high number of staples used to fasten the lath creates many penetrations that could both leak and condense moisture.
- The staples were driven into the lath with excessive force causing the lath to cut the paper creating a leak.
- Wind driven water is getting on the wall through the soffit vents and running down the wall between the sheathing and the paper.
- Weep screeds were not used at the bottom of the stucco. Lack of weep screeds can prevent trapped water from draining properly.
- Stucco was installed below ground. This may prevent trapped water from draining or may wick water up to the framing. In addition, when stucco is applied below grade there is no clear definition of where grade should be and often the grade is placed against the wood framing causing a guaranteed rot situation.
- Stucco is installed directly on foundation caps, without paper or a weep screed. This prevents trapped water from draining.
- Landscape trees or bushes that contact the stucco create an area that introduces and holds moisture in the stucco. The moisture permeates into the wall.

PAPER AND LATH REPAIR

If large areas of the existing building paper are torn or deteriorated, it should be replaced with new paper that meets the local code requirements.

- Slip the new paper behind the existing paper at the top and sides, and allow it to extend beyond the old paper at an existing overlap.
- All overlaps should be at least 4" (100 mm), to prevent water from running behind the paper.

Corroded or otherwise damaged lath should be cut out and replaced with a new section of lath conforming to ASTM C 847, ASTM C 933 or ASTM C 1032 or local building codes.

- If using paper-backed lath, separate the lath from the paper so the existing building paper is in contact with the new paper and the existing lath is in contact with the new lath.
- Overlap the paper at least 4" (100 mm), as above.
- Anchor the new lath to vertical framing members (studs) at no more than 6" (180 mm) on center.
- Avoid nailing the lath to sheathing between the studs.
- Wire-tie the new and existing lath together at least once every 6".

LATH AND PLASTER REMOVAL/PATCHING

- Stucco that has delaminated from the base, and stucco that is soft or crumbling, can usually be removed with a chisel or other hand tools.
- If the plaster is still firmly bound to the lath, the lath may have to be cut.
- Remove several inches of sound plaster around the perimeter of the area to be patched, to make room for a proper overlap of new and existing lath.
- Angle the hole in the plaster so that it is smaller toward the interior and larger toward the exterior surface. (See Figure 9-1.)
- Wire-tie the new and existing lath together at least once every 6".
- The edges of the existing stucco should be roughened and properly dampened. Bonding agents should also be used.
- Replace the stucco in layers (scratch coat, brown coat, and finish coat) that are the same thickness as the existing plaster, or with monolithically applied plaster.
- The base coat will be the smallest area to patch, and the finish coat will be the largest.

HOW-TO'S
Cut under the face of each layer to create a "key," and expose at least 2" of each layer on each side and at the top and bottom of the patch.

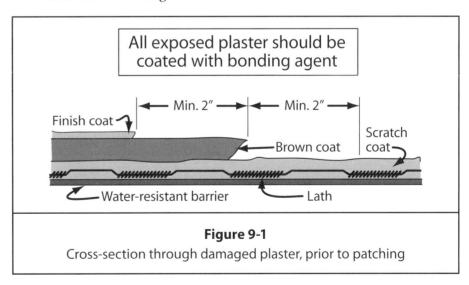

Figure 9-1
Cross-section through damaged plaster, prior to patching

- In a three-coat application, apply the scratch coat at the same thickness as the surrounding scratch coat and with enough pressure to completely embed the metal lath (when present).

- As soon as the scratch coat becomes firm, score the surface in one direction only. Vertical wall surfaces should be scored horizontally. Horizontal surfaces may be scored at any angle.
- The plaster must be moist-cured by periodically applying a fine mist of water. An easier approach is to cover the patch area completely with plastic film. Using an opaque film will protect the plaster from sunlight, although dark plastic can absorb heat and become too hot; avoid using dark garbage bags.
- Either apply the brown coat to the scratch coat as soon as it is sufficiently rigid to support additional plaster, or moist cure the scratch coat for at least 48 hours before applying the brown coat.
- Moist-cure the brown coat using the same technique as the scratch coat. Once the cure period of 2-3 days has passed, you should allow the brown coat to dry completely to equalize absorption.
- Just prior to applying the finish coat, dampen the entire surface.
- Apply the finish coat, matching the existing surface's texture. See Chapter 5.
- Color differences between patches and the surrounding stucco is unavoidable due to different mixes, age, and the effects of fading and weathering. Some of the visual contrast can be reduced by cleaning the building so the existing plaster more closely matches the freshly placed repairs.
- If you're determined to match the color and texture of the existing stucco exactly, you may find it's easier to sandblast the finish coat from the existing stucco surrounding the patch area (sandblast the entire panel). Then apply a full finish coat to the patch and surrounding area.

CRACK REPAIR

Fine cracks will develop in even the best stucco surface. Before you begin to repair cracks, you should determine their severity and probable cause, and consider the consequences if the crack is allowed to remain. Most repairs are more noticeable than the cracks they address, so most cracks should not be repaired for cosmetic reasons alone. Cracks can be the result of natural processes, or they can indicate serious problems with the structure or the stucco chemistry.

Typical causes of cracks include:
- Shifting/flexing of the base/subsurface.
- Soil subsidence, foundation settlement and framing movement.

- Alkali-silica reactivity (ASR) is an expansive reaction that occurs in wet Portland cement plaster between reactive forms of silica (found in sand and other aggregates) and potassium and sodium alkalis (found mostly in Portland cement). External sources of alkali (from soil, deicers and industrial processes) can also contribute to the ASR problem. The important thing to know is that the silica and alkali form a gel that swells as it sucks water from the surrounding wet cement. The swelling creates pressure, which combined with the artificial drying caused by the gel's extraction of water, results in expansion and cracking of the stucco. ASR conditions are often characterized by map-pattern or alligator-skin cracks. ASR can be avoided through 1) proper aggregate selection (avoiding reactive silica), 2) use of blended cements, and 3) pure, mineral-free mixing water.

To select an appropriate repair method, you should first determine whether the crack is "static" or "moving."

- Static cracks are stable. They do not provide necessary stress relief for the building, so they can be filled with a rigid material such as a plaster finish coat or elastomeric coating. They can also be repaired by following the procedures for patches described above.
- Moving cracks provide necessary stress relief for the plaster. They often result from inadequate installation or total lack of control or expansion joints, a lack of strip-lath at the corners of windows and doors, etc. Because these cracks are acting essentially as control joints, they will continue to move, and they must be allowed to open and close. If the crack is filled with a rigid material, recurrent cracking is likely. For this reason, significant moving cracks will require the installation of two-piece expansion joints that can accommodate expansion, contraction and shear.

The most common repair method for moving cracks, then, is filling with knife-grade elastomeric. The crack is widened with a saw or grinder. The resulting groove is filled with an elastomeric sealant. Cracks should usually be routed to their full depth (all the way up to but not through the lath); the width of the groove is determined by the amount of movement at the crack and the flexibility of the sealant. Consult with the sealant manufacturer to determine the proper dimensions of the sealant reservoir.

Many cracks can be effectively patched by placing a 3- to 4-inch-wide strip of masking tape along the length of the crack. Cut through the tape with a knife into the crack and squeeze the elastomeric through the slit, filling the crack. Allow the elastomeric to cure 3 to 4 hours before removing the tape to expose the remaining elastomeric scar.

- Some manufacturers offer acrylic polymer sealants marketed specifically for repairing plaster cracks. These products are often referred to as "brush grade" (for narrow cracks) and "knife grade" (for wider cracks). In addition, many manufacturers offer a variety of pigmented sealants to match the color of the plaster as closely as possible.

After the crack is repaired, the cured sealant will be glossy, which will contrast with the roughened plaster texture, making it very noticeable. One way to disguise this contrast is to throw silica sand onto the applied sealant immediately after tooling to give the sealant a roughened texture.

CLEANING PLASTER

The best way to clean stucco is to prevent it from getting dirty in the first place. Walls can become stained by rain splashing on exposed dirt alongside a stucco-finished building. This staining can be prevented by one of the following approaches:

- By planting these areas, or covering them with crushed rock, gravel or bark, you can minimize the amount of dirt that splashes onto the stucco surface. (Make sure the finish grade for landscaping is at least 4" below the weep screed.)
- In a similar fashion, concrete walkways, decks, etc. along the building's perimeter will keep stains from forming. (When installing concrete, make sure the finished slab is at least 2" below the weep screed.)
- You should also make sure that sprinklers and irrigation systems are directed away from the structure.

Another source of stains is roof runoff. Installing gutters and downspouts, and keeping them free-flowing and clear of leaves and other debris, can prevent this.

When cleaning stucco, it's always best to use the least aggressive method possible. Test any cleaning materials on a test area that is least visible before applying them on larger areas. And like stains in other materials, the sooner you address stains on stucco, the easier it will be to remove them.

In some cases, a simple washing with the garden hose will do the trick. Here is one step-by-step process that has proven itself:

- When washing the wall, pre-wet the entire wall surface, saturating it from bottom to top. (Starting at the bottom and moving up the wall prevents dirty water from the upper portions of the wall being absorbed by lower wall surfaces.)
- Now that the wall is saturated, start at the top and direct a stream of water against the wall at moderate-to-high pressure, to loosen dirt. Work your way to the bottom.

- For caked-on dirt, you may need to use a brush or broom with synthetic or natural bristles. Do not use a wire brush: the metal bristles are much too aggressive can remove the finish coat stucco.
- If greater cleaning is needed, you can try "power washing" the plaster. Keep the stream of water moving over the surface to prevent erosion of the plaster.
A fan-type spray tip that produces a 15- to 40-degree spray pattern is most effective for cleaning stucco.
Tips that produce a more concentrated stream of water may damage the surface.
- If even greater cleaning is needed, you can try a mild cleaning product.
    - Mild detergents will remove oil-based dirt and stains.
    - Cleaners based on organic solvents can remove caulking residues and asphalt materials.
    - Acids or alkalis will remove soot and other hydrocarbon residues, biological growth (mold, mildew, etc.) and stains due to air pollution.
    - Specialty cleaners (designed for stucco maintenance) are also available. Some of these will "bleach" the stain; care should be exercised, as these bleach products can also affect the color of the stucco finish or paint.
    - Keep in mind that cleaning solutions can present health and environmental hazards. Read, understand and follow all documentation supplied by the manufacturer and take all recommended precautions to maintain a safe working and living environment.
    - Flush off any remaining dirt, or any cleaning product residue, with a follow-up rinse.

In addition to these "wet" methods of cleaning stucco, there are also dry methods:
- Sandblasting
- Flame cleaning
- Shotblasting
- Grinding
- Scouring

As mentioned before, wire brushes should be used with care because they are extremely abrasive and can dig into the finish most aggressively. They can also leave metal particles behind that can themselves rust and stain the concrete.

The reason stucco lasts so long is that the cement is also mixed with sand. Silica sand (the raw material for glass) resists wear and weather for years, where a "white-wash" is just watery white cement brushed over the wall. This begins to chalk and wear within one year. To make things worse, just like painted stucco, it must be sandblasted off when a re-stucco is desired. A traditional stucco coating would consist of this same white-wash coat, however, over that another coat is applied consisting of a cement mixture including twice as much sand, thus creating strength and resistance to aging and weather. The white cement may also be colored for a pleasing look.

REFINISHING OR CHANGING THE COLOR

A stucco finish will last for years, because the materials it contains are so tough: silica sand (the raw material for glass) and cement can both resist wear and weather for years. Eventually, a stucco finish may need to be "freshened up." Or, you may just want to change the color of the building.

You have several choices to make when selecting a refinish or re-coloring method. If you simply want to stucco some damaged or new areas–e.g., around new windows or doors; on a building addition or modification; or refinishing walls that are showing signs of being exposed to harsher conditions than other areas–you may be able to find a current stucco finish coat product to match your existing finish.

If you are very fortunate, you may still have the original manufacturer's name and color number. Keep in mind, however, that your home may fade in color over time, and if your home is old you may need to ask the manufacturer to match the color of your home in their color lab.

If the objective is simply to change the color of the stucco finish, this can best be accomplished using one of the many cement-based paints that are now offered, exactly for this purpose. They are applied using brushes or rollers, but a sprayer rig is recommended. Because these products are cement-based, they form a good bond with the existing stucco surface, and won't peel or flake the way that the typical house paint can, over time.

Most cement based paints allow the exterior system to continue to "breathe", whereas some paint finishes and coatings can actually seal the stucco to which they are applied, thereby inhibiting the ability of water vapor to escape from the building interior.

For these reasons, most professionals recommend either applying a new stucco finish coat over the existing surface, or "painting" the building with one of the cement-based paints, to maintain the breatheability and durability that was provided by the original stucco finish.

Preparing the existing surface and applying the refinish product will depend on instructions and guidelines in the documentation provided by the manufacturer.

The addition of emulsified acrylic admixtures to the re-stucco finish or fog coat will enhance adhesion and improve resistance to moisture penetration.

Table 9-1
Common Problems in Stucco Installations

PROBLEM / DEFECT	LIKELY CAUSE	SOLUTION / NEXT TIME
Plaster stiffens immediately after application on concrete masonry base.	Base too dry.	Moisten base prior to plaster application, using the Saturated/Surface Dry (SSD) standard.
Hand-applied plaster falls away from metal reinforcement when applied.	Metal lath is "upside down." Plaster cohesion insufficient. Water content in the mix is too high.	Position lath properly. Adjust mix to include an air-entraining material or admixture. Reduce water content.
Scratch-coat is cracked over each stud.	Non-uniform plaster thickness.	Correct lathing procedure. Apply brown coat.
Efflorescence on scratch coat after seven days exposure.	Cold weather. Too long delay between coats.	Wash down wall with water, then apply brown coat. Heat water and sand if weather is cold.
Finish-coat is blistering.	Finish coat mix is overly "rich." Surface is over-troweled.	Adjust mix and correct your finishing technique.
Finish-coat color is uneven.	Water was dashed on surface during finishing.	Moisten basecoat plaster prior to the finish-coat application. Apply a fog coat to plaster surface to correct color.
Poor bond to concrete or concrete masonry substrate.	Base was too dry at time of application, resulting in insufficient hydration.	Moisten base to Saturated/Surface Dry (SSD) standard. Moist-cure plaster.
Poor bond to concrete or smooth concrete masonry substrate.	Substrate surface was too smooth or dense.	Select concrete masonry units with open surface texture. Use a dash bond coat, a bonding agent, or attach metal reinforcement to provide proper key.
Poor bond to concrete or masonry.	Surface covered with dirt or other contaminant.	Avoid contamination of surfaces during construction. Wash, brush clean, sandblast, or attach metal reinforcement.
Plaster cracks in craze pattern and is convex.	Brown coat has dried harder than scratch coat, or finish coat has dried harder than base coat.	Increase sand content of brown-coat plaster as compared to scratch coat. Use lower strength finish coat.
Horizontal and vertical cracks appear where metal lath overlaps.	Improper laps using paper-backed metal lath. Tie wire overlaps.	Lap paper-backed metal lath with paper-to-paper and metal-to-metal. Use sufficient number of tie wires, with recommended spacing.
Cracking	Poor consolidation—air isn't removed.	Correct rodding and floating procedures.
	Inadequate/insufficient control joints.	Strategically place and properly install/abut control joints.
	Premature moisture loss.	Moist-cure. Use plastic wrap, if necessary.

Adapted from Melander, Farny and Isberner, Portland Cement Plaster (Stucco) Manual EB049, Portland Cement Association (2003), page 35.

CHAPTER 10

Glossary

A

ACCELERATOR: An additive (admixture) that speeds the time required for one or more of the processes associated with Portland cement plaster: the rate of hydration (wetting), setting, hardening, or strength development.

ACCESSORIES: Term used to describe special lath aids such as corner beads, control joints, etc. Also used to describe those items other than frames, braces, or members used to facilitate the construction of scaffolding towers or structures.

ADDITION: A substance that is interground or blended in limited amounts into a hydraulic cement during manufacture-not at the job site-either as a processing addition to aid in manufacturing and handling the cement or as a functional addition to modify the useful properties of the cement. Improperly called additive. See: Admixture.

ADHESIVE BACKED MEMBRANE: Adhesive backed rubber sheet membranes made with tear-off paper intermediate facers. Generally used in mitigating weather barrier and flashing installations.

ADMIXTURE: A material, other than water, aggregates, and hydraulic cement, used as an ingredient of Portland cement plaster, which is added to the batch to change the performance and/or workability of the plaster. See: Accelerator, Addition, Air-entraining agent.

AESTHETIC RUSTICATION: Grooves/ reveals placed in plaster to delineate lines and shadowing effects.

AGGREGATE: A granular material such as natural sand, manufactured sand, vermiculite, or perlite.

AIR-ENTRAINING CAPACITY: The capability of a material or process to develop a system of minute bubbles of air in cement paste, mortar, or plaster during mixing.

AIR-ENTRAINING AGENT:. An addition for hydraulic cement or an admixture for plaster that will cause air to be incorporated in the form of minute bubbles in the plaster during mixing, usually to increase its workability and its frost resistance. See: Admixture.

AIR-ENTRAINING HYDRAULIC CEMENT: Hydraulic cement containing an air-entraining addition in an amount that will cause the cement to entrain air in plaster within specified limits.

AIR ENTRAINMENT: The intentional introduction of air in the form of minute, disconnected bubbles (generally smaller than 1 mm) during mixing of Portland cement plaster to improve cohesion and workability or to impart other desired characteristics to the plaster.

AMERICAN NATIONAL STANDARDS INSTITUTE, INC. (ANSI): An organization, which develops standard safety specifications and procedures for a wide range of industrial equipment, including ladders and scaffolds. Many OSHA regulations are based on these standards. Also known as ANSI.

AMERICAN SOCIETY FOR TESTING AND MATERIALS (ASTM): One of the largest voluntary standards development systems in the world, responsible for many of the standards referenced/required in the International Building Codes.

ANHYDRITE: A mineral consisting mainly of anhydrous calcium sulfate, $CaSO_3$.

ANSI: See American National Standards Institute, Inc.

APPROVED: Accepted by the authority having jurisdiction.

ATOMIZER: Device by which air is introduced into material at the nozzle to regulate the texture of the machine-applied plaster.

B

BACK PLASTERING: Applying a backup coat (or coats) of plaster to the backside of a solid plaster partition after the scratch coat has been applied and has set on the lathed side.

BASE COAT: The total of all plaster coats applied prior to application of the finish coat. The combined scratch and brown coats are the base coats.

BATCHING: Weighing or volumetrically measuring and introducing into the mixer the ingredients for a batch of plaster.

BASECOAT FLOATING: The finishing act of spreading, compacting, and smoothing of the base coat plaster to a reasonable true plane.

BEDDING COAT: The coat of plaster to receive aggregate or other decorative material of any size impinged or embedded into its surface.

BINDERS: Cementing materials, either hydrated cements or products of cement or lime and reactive siliceous materials, cement type and curing conditions govern the binder formed.

BLEEDING: The accumulation of mixing water on the surface of plaster caused by compression of the solid materials within the mass. Bleed water occurs on the surface of a mass of plaster in a container or hopper, commonly influenced by vibration. (Also called water gain.)

BLENDED HYDRAULIC CEMENT: A binder produced by intimately and uniformly intergrinding it: blending two or more types of fine materials, such as Portland cement, ground granulated blast-furnace slag, fly ash, silica fume, calcined clay, other pozzolans, and hydrated lime.

BLISTERING: Protuberances on a coat of plaster during or soon after the finishing operation; also bulging of the finish plaster coat where it separates and expands away from the base coat.

BLOCK: A concrete masonry unit, usually containing hollow cores.

BOND: Adhesion of plaster to other surfaces against which it is applied; adhesion of cement paste to aggregate; adhesion between plaster/stucco coats or between plaster and a substrate. (See "Chemical Bond" and "Mechanical Bond.")

BONDING AGENT: A compound applied a coating to a suitable substrate to enhance a bond between it and a succeeding layer, as between a subsurface and a succeeding stucco application. Also a compound used as an admixture to increase adhesion at the mortar-substrate interface and increase cohesion of the plaster

BONDBREAKER: A material used to prevent adhesion of newly placed plaster to a section of the substrate.

BOND STRENGTH: The adhesion developed 'between plaster and a substrate; the resistance to separation of plaster from other materials in contact with it.

BROWN COAT: The second coat of three-coat plastering; to complete application of a base coat.

BUILDING STRUCTURE: A wall assembly consisting of wood or steel studs with or without sheathing.

C

CALCIUM ALUMINATE CEMENT: A cement product that when added to Portland cement plaster will accelerate the set.

CALCINE, CALCINING: To make powdery or to oxidize by removing chemically combined water by action of controlled heat.

CARBONATION: Reaction between carbon dioxide and a hydroxide to form a carbonate, especially in cement paste or plaster; the reaction with calcium compounds that produces calcium carbonate.

CATFACE: Blemishes or rough depressions in finish plaster comparable to pockmarks.

CEMENT, HYDRAULIC: Any cement, such as Portland cement, that will set and harden by chemical interaction with water and that is capable of doing so underwater.

CEMENT PASTE: A mixture of hydraulic cement and water, both before and after setting and hardening.

CEMENT PLASTER: A mixture of Portland cement, Portland cement and lime, masonry cement, or plastic cement and aggregate and other approved materials as specified in the code.

CHANNEL IRON: See Cold-rolled channel.

CHECKING: Development of shallow cracks at closely spaced but irregular intervals in the plaster surface. (Also known as alligator cracking or craze cracks.)

CHEMICAL BOND: Adhesion between dissimilar materials or between one plaster coat and another that is the result of a chemical reaction.

CHOPPED FIBERS: Fiberglass or other strand materials approximately ½-inch long used in the stucco mix to provide better cohesiveness.

COAT: A layer or film, as a thickness of plaster or of paint, applied in a single operation. See: Base coat, Brown coat, Finish coat.

COHESION: The ability of a material to cling to itself.

COLD JOINT: The aesthetically objectionable and noticeable joining of fresh stucco applied next to set stucco.

COLD-ROLLED CHANNEL: A heavy-gauged, roll-formed, C-shaped metal fabrication used as the main lateral support or cross furring for a suspended plaster ceiling. Sometimes referred to as "channel iron."

COMBINED WATER: The water chemically held as water of crystallization, by the calcium sulfate dihydrate, or semihydrate crystal.

CONSISTENCY: The relative mobility or ability of freshly mixed plaster to flow.

CONTACT CEILING: A ceiling with metal lath attached in direct contact with the construction above without use of runner channels or furring.

CONTROL JOINT (CONTRACTION-EXPANSION JOINT): A single component joint with an accordion shaped profile placed in a stucco membrane that opens and closes minimally as a result of the thermal expansion and contraction and normal shrinkage of stucco. A designated separation in the plaster system that allows for movement caused by expansion or contraction of the system. The construction of the separation is accomplished by installation of manufactured devices for this application, or by fabrication procedures in the field using suitable materials.

CORNER BEAD: Metal lath accessory that provides reinforcement for plaster; used at comers to provide continuity between two intersecting plaster planes. Primarily used with gypsum plaster on interior construction.

CORNER REINFORCEMENT: Components such as corner beads and corner aids, used to plumb, strengthen, gage and provide continuity between intersecting walls that meet at an outside corner. Specially designed lath accessory with large openings to enable full plaster embedment on external corners to reinforce the plaster.

CORROSION: Disintegration or deterioration of metal reinforcement due to electrolysis or chemical attack.

CRAZE CRACKS: Fine, random cracks or fissures that may appear in a plaster surface, caused by shrinkage. *See* Checking.

CROSS FURRING: Roll-formed channel attached perpendicular to the main runners for the attachment of metal lath in a suspended plaster ceiling.

CROSS SCRATCHING: Scratching of the first coat in two directions to provide a mechanical bond between coats. Not recommended for Portland cement plaster scratch coat.

CURING: The process by which stucco ultimately reaches its full hardness and strength, usually involving keeping freshly applied plaster moist and at a favorable temperature for a suitable period following application. Curing assures satisfactory hydration and carbonation of the cementitious materials and proper hardening of the plaster.

CURLING: The distortion or warping of an essentially planar surface into a curved shape, due to several factors such as temperature and moisture differences within the plaster coat.

D

DAMPPROOFING: Treatment of plaster to retard the passage or absorption of water, or water vapor, either by application of a suitable surface treatment on exposed surfaces or by use of a suitable admixture or treated cement.

DARBY: A straight-edged tool with handles, approximately 42-inches-long, made of magnesium alloy or wood. Used to float and smooth the fresh stucco brown coat.

DASH-BOND COAT: A thick slurry of Portland cement and fine sand plus sufficient water that is dashed by hand or machine onto concrete, masonry, or older plaster surfaces to provide a mechanical bond for succeeding plaster.

DASH COAT: A wet plaster coat splatter-applied to a surface as a final finish texture. When applied or to provide a mechanical key for a subsequent application of a brown coat, it is typically called a Dash-bond coat.

DASH TEXTURE: A finish coat of thick plaster hand-dashed or machine applied onto a well prepared, uniformly plane surface of brown-coat plaster. (Also called spatter-dash or splatter-dash.)

DECORATOR'S PLANK: Fabricated planks of both the extension and fixed length types used for supporting one worker and limited material. These planks are usually used with ladder jack scaffolds, trestle, extension trestle, platform and step ladders.

DEFLECTION: The limits at which an applied axial or wind load on a structural member will cause damage in a stucco membrane. The design deflection criteria for stucco should be L/360; that is, the length of the span (in inches) divided by 360.

DISCOLORATION: Change in color from the normal or desired.

DOUBLE-BACK: The process of installing the brown coat immediately after the scratch coat has reached sufficient rigidity to accept it. (Also called double-up.)

DRIP SCREED: A device used to interrupt the flow of water on a wall. A weep screed would be an example of this.

DURABILITY: The ability of Portland cement plaster to resist weathering action, chemical attack, abrasion, and other potentially harmful service conditions.

E

EARTH PIGMENTS: The class of pigments produced by physical processing of materials mined directly from the earth; also frequently called natural or mineral pigments or colors.

EFFLORESCENCE: A deposit of salts or bases, usually white, formed on a surface. Water soluble substances emerge in solution from within the plaster and are deposited during evaporation.

EIFS: *See* Exterior Insulation and Finishing System.

ENGLISH TEXTURE: A texture created by using a rounded trowel to apply final texture in a random pattern.

EPS: *See* Expanded polystyrene.

EXPANDED METAL LATH: Sheets of metal that are slit and pulled out to form diamond-shaped openings; used as metal reinforcement for plaster.

EXPANDED POLYSTYRENE (EPS): Often referred to using the Dow Chemical Company trade name Styrofoam®. Used in panels as sheathing in EIFS, and for decorative plant-ons to achieve architectural effects under stucco.

EXPANSION JOINT: Generally refers to a telescoping two- or three-piece accessory designed to control structural or seismic stresses. *See* Control joint.

EXTERIOR INSULATION AND FINISHING SYSTEM (EIFS): Proprietary cladding system incorporating an insulation board and an acrylic finish coat.

F

FACTORY PREPARED STUCCO MATERIALS (MILL-MIXED OR READY-MIXED): Pertaining to material combinations that have been formulated and dry blended by the manufacturer, requiring only the addition of and mixing with water to produce stucco (Stucco finish coat)

FACTOR OF SAFETY (SCAFFOLDING AND SHORING): The ratio of ultimate load and the allowable load.

FALL PREVENTION DEVICE: Physical components used to prevent a worker from falling to a lower surface. These include but are not limited to guardrails, screens, body belts, body harnesses, tethers, etc.

FAUX FINISH: Generally refers to a painted surface made to imitate a variegated plaster finish (French).

FEATHEREDGE: A wood or metal tool with a beveled edge and varying in length; used to straighten reentrant angles in finish-coat plaster. Also used to plane the surface of the brown coat and dry rod or dry rake it to better control color in the finish coat.

FIBERS: Natural or synthetic. An elongated fiber or strand admixture added to stucco mix to improve cohesiveness, strength, and helps reduce cracking.

FINE AGGREGATE: Natural or manufactured sand that passes the No. 4 (4.75 mm) sieve.

FINISH COAT: The final coat of plaster, the decorative surface, usually colored and textured.

FLASHING: Approved, corrosion-resistive material provided in such a manner as to deflect and resist entry of water into the construction assembly. A thin, non-vapor permeable material used to prevent water entry and/or direct the water migration in a desired direction between two or more materials and/or surfaces.

FLOAT: *(noun)* A rectangular tool consisting of a handle attached to a base pad of molded rubber, foam plastic, cork, wood, or felt tacked to wood and used to impart a relatively even but still open texture to a plaster surface. *(verb)* The act of smoothing/leveling a surface, using a float.

FLOATING: The process of compacting and leveling the stucco basecoat.

FOG COAT: A fine mist of cement based paint color used to provide uniformity in integral colored cement finish coat.

FRIEZE FINISH: The finish coat is brushed (broomed) prior to final application of a trowel texture.

G

GRADATION: The size distribution of aggregate particles, determined by separation with standard screen sieves.

GROUND: A piece of wood or metal attached to the framing with its exposed surfaces acting as a gauge to determine the thickness of plaster to be applied. That part of an accessory which establishes the thickness of the stucco and also mechanically keys to the stucco.

GROUT: Portland cement plaster mortar used to fill crevices or to fill hollow metal frames.

H

HAIRLINE CRACKS: Very fine cracks in either random or essentially straight line patterns that are just visible to the naked eye. (See "Checking.")

HANGER: Wire, threaded rod or metal strapping used to support main runners to the construction in a suspended plaster ceiling installation.

HARSH MIXTURE: A mixture that lacks desired consistency and workability due to a deficiency of cement paste, aggregate fines, sufficient water, or a combination thereof.

HAWK: A tool with a flat metal surface supported by a single handle that is used to hold plaster before it is transferred to a trowel for application.

HOIST: A mechanical device used to raise or lower a suspended scaffold It can be either manually or mechanically power operated.

HYDRATED LIME: The product manufactured by heating limestone until carbon dioxide is removed, thus forming quicklime (calcium and magnesium oxides), subsequently hydrated using water additions. Hydrated lime processing involves pressure hydration, atmospheric hydration, or slaking.

HYDRAULIC CEMENT: See "Cement, hydraulic."

I

IMPLANT: *See* Plant-on.

L

LACE FINISH: Refers to a lighter skip trowel finish than Spanish texture.

LATH: Reinforcement base to receive plaster, generally secured to substrate. The expanded metal, welded wire or woven wire that is attached to a building's structural elements, acting as an armature for the adhesion and mechanical key of the stucco.

M

MAP CRACKING: See "Craze cracks." Mechanical bond. The physical keying of one plaster coat to a substrate or another coat; or plaster keying to metal lath; or interlock between adjacent plaster coats created by surface irregularities, such as 'scratching.

MAIN RUNNER: See Cold-rolled channel.

MECHANICAL KEY: The process of roughening a surface for the subsequent application of plaster. See also Scarifying, Scratch coat and Dash-bond coat.

METAL LATH: Metal lath is slit and expanded or stamp punched from plain or galvanized steel coils or sheets. It is of two types: diamond mesh lath, which may be flat or self furred with impressed indentations, and rib lath. Metal lath is coated with a rust-inhibiting paint after fabrication or is galvanized.

MOIST CURING: Any method employed to retain sufficient moisture for hydration of Portland cement plaster.

MOISTURE MOVEMENT: The migration of moisture within and from a porous medium, caused by an imbalance as surface moisture is lost through evaporation.

MORTAR: A mixture of hydraulic cement, water, fine aggregate, and possibly plasticizers.

N

NOZZLE: Attachment at discharge end of delivery hose used for machine application of plaster. The nozzle allows adjustment of the spray pattern.

NOZZLEMAN: Person who manipulates the nozzle and controls plaster placement.

O

OCCUPATIONAL SAFETY AND HEALTH ACT (OSHA): A federal law setting forth rules for the safety working practices and equipment in all areas of business and industry. Also known as OSHA.

ONE-COAT STUCCO: A proprietary manufactured stucco base coat product that is installed in one thin application pass.

P

PARGE: An archaic plaster term used to describe coating masonry with plaster.

PLANK: A wood member or fabricated component which is used to create the horizontal working surface of a scaffold system.

PLANT-ON: A shape or form made of formed expanded polystyrene (EPS), or similar material, that is glued to the plaster or sheathing, and then covered with a glass fiber mesh embedded in a thin coat of colored stucco, to achieve architectural effects such as arches, column details, etc. *Also called* Implant.

PLASTER GROUND: A piece of wood or trim accessory that acts as a termination and/or gauge to determine the thickness of the plaster.

PLASTER THICKNESS: See "Thickness~ plaster."

PLASTICITY: A complex property of plaster involving flow of the plaster associated with an applied force; that property or freshly mixed plaster that determines its resistance to deformation or its ease of molding.

PLASTICIZER: An additive that increases the plasticity of a Portland cement plaster. Plasticizing agents include hydrated lime or lime putty, air entraining agents, organic additions and fine ground or processed inorganic substances.

PLUMB: Exactly straight up and down; vertical.

PORTLAND CEMENT PLASTER: A mixture of Portland cement or Portland cement and lime and aggregate and other approved materials as specified in the code.

PVC: Poly Vinyl Chloride. Plastic that is used to manufacture stucco trim.

R

RAIN SCREEN: Generally refers to an exterior wall cladding system with a measurable airspace between the building sheathing material and the backside of the exterior cladding system.

REENTRANT CRACK: A hairline crack that can develop at a natural stress point (e.g., the apex of a 90 degree corner, at the corner of an opening, or a penetration such as at a window or door).

REGISTERED PROFESSIONAL ENGINEER: A person who has been duly registered and licensed by an authority within the United States to practice the profession of engineering.

RETARDATION: Slowing down the rate of hardening or setting of plaster to increase the plaster setting and hardening times.

RE-TEMPERING: After initial mixing, adding water and remixing plaster that has started to stiffen and become harsh.

REVEALS: Aesthetic grooves or rustication placed in plaster to delineate lines and shadowing effects.

ROD: A straightedge (usually more than 5′ long), made of magnesium alloy or wood, used to "rough plumb" (i.e., straighten the face of walls and ceilings by cutting off excess plaster to a plane established by forms, ground wires, or screeds).

RODDING: A method of straightening the stucco basecoat.

S

SAND: See "Fine aggregate."

SAND FLOAT FINISH: A final texture of finish coat, also known as "sand" finish. The texture will vary depending on the sieve aggregate size.

SANTA BARBARA OR MISSION FINISH: Generally refers to a semi smooth trowel finish with blemishes (catfaces) in the final troweling that may include a wavy brown coat.

SCARIFIER, OR SCRATCHER: A tool with flexible steel tines used to scratch or rake the unset surface of a first (scratch) coat of plaster.

SCARIFYING: The process of raking a fresh plaster surface to provide a mechanical key for the subsequent application of an additional coat of plaster.

SCORING: Grooving by scratching or scoring, usually horizontal, of the scratch coat to provide mechanical keys with the brown coat.

SCRATCH COAT: First coat of plaster applied to a surface in two- or three-coat plastering work, that is then scarified to create a mechanical key for the subsequent brown coat.

SCREED: A tool used to establish a flat plane and uniform stucco finish. An accessory or component that aids in gauging the thickness of the stucco. Casing beads, weep screeds, control joints and corner beads would all qualify as screeding devices.

SELF-FURRING: Metal lath or wire lath formed during manufacture to include raised portions of the lath, ribs, or dimples that hold the lath away from the supporting surface and position it for embedment with plaster.

SET: The change in plaster from a plastic, workable state to a solid, rigid state. Set is modified by the terms "initial" and "final," both arbitrary appraisals of degree of hardening.

SKIM COAT: A finish coat applied to an existing stucco surface or other substrate to improve appearance.

SLICKER: A wood or metal straight-edge tool used in lieu of a darby to float and smooth a stucco surface to plumb.

SPANISH TEXTURE: Refers to a medium to heavy skip-trowel finish coat.

SPATTERDASH: See "Dash texture."

STUCCO FINISH COAT: The final layer of stucco (job-site mixed or pre-manufactured applied over a basecoat stucco or direct to concrete.

SUCTION: The absorptive capacity of a substrate or plaster surface immediately after being subjected to application of water or plaster.

T

THICKNESS, PLASTER: Plaster thickness is measured from the back plane of metal reinforcement or from the face of solid backing or support to the specified plaster surface, either scratch, base, or finish. Sometimes referred to as "grounds."

TIE WIRE: Annealed soft temper steel wire used for a variety of lathing operations (e.g., to attach lath to accessories).

TRIM ACCESSORIES: Components installed during the lathing installation such as casing beads, control joints, weep screeds, etc.

TROWEL: A flat, broad-blade steel hand tool used to apply, spread, shape, and smooth finish-coat plaster.

U

UNDERWRITERS LABORATORIES, INC.: This organization tests the safety aspect of a wide variety of equipment used in industry and in the home, in accordance with a set of engineering and design standards. Also known as UL.

V

VAPOR PERMEABLE: A material that will allow water to pass through as a gas (vapor). Vapor permeable materials are rated in terms of the permeance unit, *perm*. This unit is defined as a vapor transmission rate of one grain of water vapor through one square foot of material per hour, when the vapor pressure difference is equal to one inch of mercury. A material having a vapor transmission rate of one perm or more as a property of the substance is designated as "vapor permeable."

VARIEGATED: Finish surface which is irregularly marked with different colors.

VENETIAN PLASTER: Generally refers to a finish coat made from marble dust and lime, which may be troweled smooth to imitate marble or a variety of other finishes.

W

WARPING: A deviation of a wall surface from its original shape, usually caused by temperature or moisture differentials within the plaster. Also caused by an excessively rich (high-strength) finish-coat plaster or by hard troweling to produce a smooth finish-coat surface. (See also "Curling.")

WATER-RESISTANT OR WEATHER-RESISTIVE BARRIER: Correct terminology is water-resistive barrier, however, verbiage has not yet caught up in all circles. These barriers are any of the variety of housewraps, building papers or felts that have been available to wrap the external sheathing of a building prior to the installation of the stucco or other cladding system.

WIRE MESH LATH: Plaster reinforcement available in two types, woven wire and welded wire. Woven wire is made of galvanized wire woven or twisted to form either squares or hexagons. Welded wire is zinc coated and electrically welded at all intersections. Both types may be paper backed, and are available in rolls or sheets.

WORKABILITY: The property of freshly mixed plaster that determines its working characteristics, i.e., the ease with which it can be mixed, placed, and finished.

Z

ZINC ALLOY: A soft malleable bluish-white metal, highly resistant to oxidation.

CHAPTER 11

Associations and Major Manufacturers

MANUFACTURING AND TRADE ASSOCIATIONS

American Concrete Institute (ACI)
Box 19150, Redford Station
Detroit, MI 48219
www.aci.org

ASTM International
American Society for Testing and Materials
100 Bar Harbor Drive
P.O. Box C700
West Conshohocken, PA 19428-2959

ASTM is one of the largest voluntary standards development systems in the world. Many ASTM standards are referenced and required in current International Building Codes.

Association of the Wall and Ceiling Industry (AWCI)
803 West Broad Street, Suite 600
Falls Church, VA 22046
(703) 534-8300
www.awci.org

AWCI represents acoustics systems, ceiling systems, drywall systems, exterior insulation and finishing systems, fireproofing, flooring systems, insulation, and stucco contractors, suppliers and manufacturers and those in allied trades.

Carolina Lathing and Plastering Contractors Assn.
P. O. Box 7582
Charlotte, NC 28241
www.clapca.org

EIFS Industry Members Association
3000 Corporate Center Drive, Suite 270
Morrow, GA 30260
(800)294-3462 / (770) 968-7945
www.eima.com

National non-profit technical trade association comprised of more than 400 leading manufacturers, suppliers, distributors and applicators involved in the exterior insulation and finish systems (EIFS) industry.

EPS Molders Association
1298 Cronson Boulevard, Suite 201
Crofton, MD 21114
(800) 607-3772
www.epsmolders.org

Lath and Plaster Institute of Northern California
1043 Stuart Street, #2
Lafayette, CA 94549

The Minnesota Lath & Plaster Bureau
820 Transfer Road
St. Paul, MN 55114
(651) 645-0208
www.mnlath-plaster.com

The Minnesota Lath and Plaster Bureau has promoted the industry since 1953. It is widely recognized as an education and technical spokesman for the industry. It provides services to architects, the construction community and the public on a variety of matters relating to the plastering trades. It's mission is to promote the quality, pride and professional craftsmanship of Minnesota's union plastering contractors through communication, education and information.

Northwest Wall & Ceiling Bureau (NWCB)
NW Wall and Ceiling Bureau
1032-A NE 65th Street
Seattle, WA 98115
(206) 524-4243
www.nwcb.org

The Northwest Wall & Ceiling Bureau (NWCB), based in Seattle (Washington) is an international professional trade association serving a wide ranging membership of contractors, manufacturers, dealers (suppliers and distributors) and labor. Their members are involved in stucco, exterior insulation finishing systems (EIFS), acoustical ceilings, gypsum wallboard (drywall), interior plaster, light-gauge steel framing, and spray-on fireproofing. Their membership is primarily located in the U.S. Northwest and Western Canada. Their activities and publications focus on the special needs of their membership, given the climate of the region.

Operative Plasterers' and Cement Masons' International Assn. (OPCMIA)
14405 Laurel Place, Suite 300
Laurel, Maryland 20707
(301) 470-4200
www.opcmia.org

The Operative Plasterers' and Cement Masons' International Association (OPCMIA) of the United States and Canada is a union that represents plas-

terers and cement masons in the construction industry in North America. Union members finish interior walls and ceilings of buildings and apply plaster on masonry, metal, and wire lath or gypsum. Cement masons are responsible for all concrete construction, including pouring and finishing of slabs, steps, wall tops, curbs and gutters, sidewalks, paving and other concrete construction.

Portland Cement Association
5420 Old Orchard Road
Skokie, IL 60077
(847) 966-6200
www.cement.org

The Portland Cement Association (PCA) membership is several dozen of North America's largest manufacturers of Portland cement products. It conducts market development, engineering, research, education, and public affairs programs.

Western Wall and Ceiling Contractors Association
Technical Services and Information Bureau
1910 N. Lime Street
Orange, CA 92865
(714) 221-5520
www.wwcca.org

WWCCA is a non-profit organization representing over 60 subcontractors and 80 affiliates who have joined to promote the installation of quality acoustical tile, decorative and ornamental plastering, drywall, EIFS, fireproofing, insulation, lathing, plastering, steel stud framing. They employ only union trained craftsmen.

Texas Lathing and Plastering Contractors Association (TLPCA)
1615 W. Abram, Suite 101
Arlington, TX 76013
(817) 461-0676
www.tlpca.org

PORTLAND CEMENT MANUFACTURERS

Bonsal American
8201 Arrowridge Blvd.
Charlotte, NC 28273
(704) 525-1621
www.bonsal.com

California Portland Cement Company
Producing Colton Aggregates, California Portland Cement, Arizona Portland Cement, and other brands
2025 East Financial Way, Ste. 200
Glendora, CA 91741-4692
(626) 852-6200
www.calportland.com

LaFarge North America
12950 Worldgate Drive, Suite 500
Herndon, VA 20170
(703) 480-3600
www.lafargenorthamerica.com

Hanson Permanente Cement
8505 Freeport Parkway
Irving, TX 75063
(972) 621-0345
www.hanson.biz

Rinker Materials, Inc.
1501 Belvedere Road
West Palm Beach, FL 33406
(800) 226-5521
www.rinker.com

TXI Riverside Cement
1500 Rubidoux Blvd.
Riverside, CA 92509
(909)774-2500
www.txi.com

Sacramento Stucco
860 Riske Lane
West Sacramento, CA 95691
(916) 372-7442
www.sacstucco.com

South Down, Inc. (Medusa)
Medusa Cement Company
2720 S Highway 341
Hawkinsville, GA 31036
(478) 987-2373

LATH / PAPER-BACKED LATH / LATH ACCESSORY MANUFACTURERS

AMICO
(Alabama Metal Industries Corporation)
3245 Fayette Ave.
P.O. Box 3928
Birmingham, AL 35208
(205) 787-2611 / (800) 366-2642
www.amico-lath.com

Bailey Metal Products
One Caldari Road
Concord, Ontario L4K 3Z9
(905) 738 6738 / (800) 668 2154
www.bmp-group.com

California Expanded Metals (CEMCO)
263 N. Covina Lane
City of Industry, CA 91744
(800) 775-2362
www.cemcosteel.com

Clark/Western
101 Clark Blvd.
Middletown, OH 45044
(800) 543-7140
www.clarksteel.com

Davis Wire
5555 Irwindale Avenue
Irwindale, CA 91706-2070
(800) 350-7851
www.daviswire.com

Dietrich Metal Framing
500 Grant Street, #2226
Pittsburgh, PA 15219
(412) 281-2805
www.dietrichindustries.com

Jaenson Wire Products

see Davis Wire

K-Lath (Division of Tree Island)
13470 Philadelphia Avenue
Fontana, CA 92337
(951)-360-8288
www.klath.com

Marino Ware
400 Metuchen Road
South Plainfield, NJ 07080
(800) MARINO-1
www.marinoware.com

Structa Wire Corp.
1395 North Grandview Hwy,
Vancouver, BC V5N 1N2
25933 Table Meadow Road
Auburn, CA 95602
www.structawire.com

Stockton Products Co.
4675 Vandenburg Dr.
N. Las Vegas, NV 89031
(877) 862-5866
www.stocktonproducts.com

TOOLS AND MACHINERY MANUFACTURERS

Goldblatt Tools
Stanley Tools Product Group
480 Myrtle Street
New Britain, CT 06053
www.stanleytools.com

Marshalltown
104 South 8th Avenue
Marshalltown, Iowa 50158
(641) 753-0127
www.marshalltown.com

Putzmeister America
1733 90th Street
Sturtevant, WI 53177
(800) 884-7210
www.putzmeister.com

Quikstir, Inc. (QuikSpray Pumps)
2105 W. Lakeshore Dr.
Port Clinton, OH 43452
(419) 732-2611
www.quikspray.com

EIFS AND MOLDED FOAM MANUFACTURERS/SOURCES

Advance Foam Plastics, Inc.
5250 N. Sherman St.
Denver, CO 80216
(303) 297-3844
www.afprcontrol.com

Dryvit Systems, Inc.
One Energy Way
PO Box 1014
West Warwick, RI 02893
(800) 556-7752 / (401) 822-4100
www.dryvit.com

Firestone Building Products Co.
A Division of BFS Diversified Products, LLC
310 E. 96th Street
Indianapolis, IN 46240
(800) 428-4442
www.firestonebpco.com

Foam Concepts (makers of Styro-Loc)
4750 E. Wesley Dr.
Anaheim, CA 92807
(714) 693-1037

www.styro-loc.com

La Habra Stucco (part of Parex Group)
P.O. Box 17866
4125 E. LaPalma, Suite 250
Anaheim, CA 92807
(714) 778-2266
www.lahabrastucco.com

Omega Products International
1681 California Avenue
Corona, CA 92881
(800) 600-6634
www.omega-products.com

ParexLahabra, Inc.
4125 E. LaPalma Ave., Suite 250
Anaheim, CA 92807
(714) 778-2266
www.parexlahabra.com

Senergy, LLC
BASF Wall Systems, Inc.
3550 St. John's Bluff Road, South
Jacksonville, FL 32224-2614
(800) 221·9255
www.senergy.cc

Sto Corp.
3800 Camp Creek Parkway
Building 1400, Suite 120
Atlanta, Georgia 30331
(800) 221-2397
www.stocorp.com

Stuc-O-Flex International, Inc.
17639 NE 67th Court
Redmond, WA 98052
1-800-305-1045 / (425) 885-5085
www.stucoflex.com

GENERAL INFORMATION WEBSITES AND SOURCES

International Institute for Lath and Plaster
P.O. Box 3922
Palm Desert, CA 92261-3922
(760) 837-9094
www.iilp.org

CHAPTER 12

References / Bibliography

ACI Committee 524 Report. *Guide to Portland Cement Plastering*, ACI 524R-93, American Concrete Institute. Farmington Hills, Michigan. 1993.

Association of the Wall and Ceiling Industry, *Technical Manual No. 15. Evaluation of Three-Coat Portland-Cement Plaster (Stucco)*.

American Society for Testing and Materials, West Conshohocken, PA.

ASTM C 79: *Standard Specification for Treated Core and Nontreated Core Gypsum Sheathing Board*.

ASTM C 91: *Specification for Masonry Cement*.

ASTM C 150: *Specification for Portland Cement*.

ASTM C 206: *Specification for Finishing Hydrated Lime*.

ASTM C 207: *Specification for Hydrated Lime for Masonry Purposes*.

ASTM C 208: *Standard Specification for Cellulosic Fiber Insulating Board*.

ASTM C 260: *Specification for Air Entraining-Admixtures for Concrete*.

ASTM C 494: *Specification for Chemical Admixtures for Concrete*.

ASTM C 595: *Specification for Blended Hydraulic Cements*.

ASTM C 645: *Specification for Nonstructural Steel Framing Members*.

ASTM C 754: *Specification for Installation of Steel Framing Members to Receive Screw-Attached Gypsum Panel Products*.

ASTM C 841: *Specification for Installation of Interior Lathing and Furring*.

ASTM C 847-95. *Specification for Metal Lath*.

ASTM C 897: *Specification for Aggregate for Job-Mixed Portland Cement-Based Plasters*.

ASTM C 926: *Specification for Application of Portland Cement-Based Plaster.*

ASTM C 926-98a. *Standard Specification for Application of Portland Cement-Based Plaster.*

ASTM C 932-98a. *Specification for Surface-Applied Bonding Agents for Exterior Plastering.*

ASTM C 933: *Specification for Welded Wire Lath.*

ASTM C 954: *Specification for Steel Drill Screws for the Application of Gypsum Panel Products or Metal Plaster Bases to Steel Studs from 0.033 in. to 0.112 inch Thickness.*

ASTM C 955: *Load-Bearing (Transverse and Axial) Steel Studs. Runners (Tracks). and Bracing or Bridging for Screw Application of Gypsum Panel Products and Metal Plaster Bases.*

ASTM C 979: *Specification for Pigments for Integrally Colored Concrete.*

ASTM C 1002: *Specification for Steel Self-Piercing Tapping Screws for the Application of Gypsum Panel Products or Metal Plaster Base to Wood Studs or Steel Studs.*

ASTM C 1007: *Specification for Installation of Load Bearing (Transverse and Axial) Steel Studs and Related Accessories.*

ASTM C 1032: *Specification for Woven Wire Plaster Base.*

ASTM C 1063-99, *Standard Specification for Installation of Lathing and Furring to Receive Interior and Exterior Portland Cement-Based Plaster.*

ASTM C 1177: *Standard Specification for Glass Mat Gypsum Substrate for Use as Sheathing.*

ASTM C 1278: *Standard Specification for Fiber-Reinforced Gypsum Panel.*

ASTM C 1328: *Specification for Plastic (Stucco) Cement.*

ASTM E 1857-97. *Standard Guide for Selection of Cleaning Techniques for Masonry; Concrete, and Plaster Surfaces.*

ASTM E 2266-04 *Standard Guide for Design and Construction of Low-Rise Frame Building Wall Systems to Resist Water Intrusion.*

Bucholtz, John J., Plaster Information Center, San Jose, CA.

Water Leaks and Water Traps in Stucco Buildings.

The Consumer's Stucco Handbook (1995).

Questions about Stucco.

Techniques and Comments.

Building Standards Institute, *California Building Performance Guidelines for Residential Construction*, Sacramento, CA.

California Building Code: 2006.

Californiia Business and Professions Code: *C-35 Plastering Contractor.*

California Department of Education, *Plastering Workbook*, Part 2. Sacramento, CA, 1974.

Cement Plaster Construction. Technical Report SL-84-10. U.S. Army Corps of Engineers. June 1984.

Craftsman Book Company,, *National Construction Estimating,* CA 2001

Gorman, J.R., and Pruter, Walt, *Plaster and Drywall System Manual, 3rd Edition,* Information Bureau, Lath, Plaster and Drywall.

Grimmer. A.. *The Preservation and Repair of Historic Stucco.*U.S. Department of the Interior, National Park Service. Preservation Assistance Division. October 1990.

International Code Council, *AC 11 Acceptance Criteria for Cementitious Exterior Wall Coatings.*

Lucente, Paul, *Study Guide for the Contractor's State Licewnse C-35 Plastering,* Builder's Book Inc.

Maylon, Gary J. *The Metal Lath Handbook.*

Melander, Farny, Isberner, *Portland Cement Plaster/Stucco Manual (Fifth Ed.),* Portland Cement Association, Skokie, IL, 2003.

Minnesota Lath & Plaster Bureau, *Stucco in Residential Construction:* www.mnlath-plaster.com.

Monash University, *Standard Specifications.*

Morgan, Morris Hickt, *Vitruvius The Ten Books of Architecture,* Dover Publishing Co.., NY 1960.

NAAM Standard ML/ SFA 920, *Guide Specifications for Metal Lathing and Furring.* Metal Lath/ Steel Framing Association: National Association of Architectural Metal Manufacturers.

Northwest Wall & Ceiling Bureau, *Portland Cement Plaster - Stucco Resource Guide (3rd Ed.),* Seattle, WA, 2002.

Operative Plasters and Cement Mason International Association, *Mission Statement.*

PCA. *Portland Cement Plaster (Stucco) Manual,* EB049, Portland Cement Association, 1996, 50 pages.

PCA. *Masonry Information - Repair of Portland Cement Plaster (Stucco)* Portland Cement Association, 2001.

Pruter, Walter F. "Crack Control in Portland Cement Plaster," *Construction Specifier,* April 2005, page 101.

Putzmeister America, *Complete Concrete Placing Systems and Support.*

Richardson Engineering Services. *Process Plant Construction, AZ 2001.*

Schwartz, Max, *Machines Buildings Weaponry of Biblical Times,* Revell, NJ, 1990.

Schwartz, Max, *Exploring Machines Buildings Weaponry of Biblical Times,* World Bibles, Wm. B. Eerdmans Publishing Co., MI, NJ, 1990.

Shreve, N. Norris, *Selected Process Industries,* McGraw-Hill book Co,, NY, 1950.

Stucco Manufacturers Association, *Stucco Textures and Finishes.*

Ribar, J. W., and Scanlon, J. M.. *How to Avoid Deficiencies in Portland Cement Plaster.*

U.S. Department Of Labor, *Plasterers and Stucco Masons.*

Van Den Branden, Hartsell, *Plastering Skills,* American Technical Publishers, Inc., Homewood, IL, 1984.

Waddell, Joseph J., *Construction Materials – Ready-Reference Manual,* McGraw-Hill Book Co., 1985.

Zwayer. G. L., "Metal Lath Placement and Stucco Cracks," The Construction Specifier (magazine). January 1999. page 72.

APPENDIX A

History of Stucco

HISTORY

Cement plaster is one of the oldest and long-lasting of construction materials. Its raw material is available in most parts of the world, and it's relatively simple to manufacture. In fact, if prepared properly, it is as permanent as the rock from which it is made. The art of plastering is as old as civilized man.

Early civilizations used plaster to protect their sun-dried brick walls, which would otherwise have dissolved in the rain. They placed cement plaster over compacted soil to serve as sturdy floors for their homes and temples. They sheathed their wood-framed roofs to protect the interior from rain, heat of the sun, and to provide additional space for cooking. Also, they sloped their flat roofs collect and drain rainfall to underground cisterns.

In the sun-parched Middle East, people dug cisterns near their houses for the storage of water from the occasional rains. They plastered the walls of these cisterns to contain the stored water.

Because of the long life of plaster, present day archaeologists found this material in the lining of public baths, aqueducts and drainage sewers. They also found building walls, floors and roofs coated with cement plaster. Sometimes, these surfaces were decorated with marble chips and color pigments to create beautiful mosaics and *frescoes*. In addition to all the properties described above, the ancient artisans used plaster to resist fire and used it to coat the inside of their ovens, furnaces and kilns.

Figure A-1
Old stucco building
Prague, Czech Republic

RAW MATERIALS

Basically, cement plaster was made of ground limestone, sand, and water, all plentiful in the ancient world. There were other types of plaster, such as gypsum, that also resisted fire but dissolved when wet. Therefore, gypsum-based plasters were used in the interior of their buildings and tombs. *Gypsum* was the Greek word for plaster.

ANCIENT MANUFACTURE

The early builders excavated limestone from quarries with hammers and chisels. Then, they crushed the rock into smaller pieces, which they placed in a wood or charcoal fire. With enough heat and time, the calcined chunks of limestone changed to an anhydrous material by losing its "chemical water." These "clinkers" were then ground to a fine powder with hard stones to form the finished material, called cement.

Sand, usually found at riverbeds or at the beach, was washed to remove any dirt or impurities. This formed the aggregate for the plaster. Sometimes a small amount of clay was added to improve the workability of the mix.

PLASTER

Finally, the early plasterers added just enough water to start hydration of the cement. They then mixed the components thoroughly with additional water. The cement completely coats the particles of sand. The calcined cement picked up the water molecules it lost while heated and began to solidify while binding the particles of sand together. After a while the plaster mix is cured and stable. The cement had changed back to limestone with grains of sand intermixed. This operation is very similar to the making of concrete except that crushed rock or gravel is added to the mix.

The modern method of manufacture is machine-driven. Quarrying is done with powered excavators, jack-hammers, and bulldozers. Material is moved by conveyor belt, bucket elevators or air-blown through pipes..

The limestone ore is broken to smaller size by crushers, then fed to long tubular steel kilns, where it is cooked by gas and oil fires. The calcined or roasted clinkers are then ground to a powder by a series of mills and screens to a fine powder, which is mixed with various additives. The final mix, called Portland cement, is stored in silos and bins, then bagged for shipment to the jobsite.

There are various types of Portland cement, depending upon the type of limestone used and additives. The additives may be lime, gypsum, vermiculite, plasticizers and other ingredients. Five basic types of cement are produced depending upon its final use and environment.

Plastering is generally divided into three stages or phases: base or scratch coat, second or brown coat and stucco or finish coat. Each coat serves a different and specific service.

- The base coat must bond to the substrate well, whether it is concrete or masonry, wood or steel framing. With framing a waterproof paper and metal lath are required for bonding.
- The brown coat has a rough finish to provides a good base for the finish coat. Hence it is called "scratch coat."
- The finish coat is artistic part of the job. It may be colored, textured or modeled. The texture includes everything from a relatively smooth sandy finish to a rough splattered finish looking like an adobe building.

Cement plaster has been used for thousands of years because of its useful properties. It is permanent as rocks it came from. It has shear strength and resists earthquake forces, It protects a wood or steel frame because it is fireproof. It is waterproof and won't dissolve when wet. When used in conjunction with waterproof paper, it protects the interior of a building from rain. And finally, it reduces sound and noise.

The modern plasterer works in conjunction with other trades. These are brick masons, wood and metal framers, sheetmetal craftsmen, lathers, and others. He uses a great variety of tools, some of which were used in biblical times. Modern tools include plaster mixers and pumps.

Their work is highly regulated by standards and codes. These include industry standards, safety standards and architect specifications. These rules provide safety to the occupants from fire, rain, earthquake, wind, and age. They also provide safety to the workmen on the job.

Finally, this book describes common plastering failures and their remedies. Also included are cost estimating methods and a Plastering Glossary.

PLASTERERS AND STUCCO MASONS

Here are some famous definitions of the plasterer:

"Plastering is both a craft and an art."

"A man who works with his hands is a laborer."

"A man who works with his and his head is a craftsman."

"A man who works with his hands, his head and his heart is an artist."

Source: Van den Branden and Hartsell, Plastering Skills, AIP Publications

Other construction workers who use a trowel as their primary tool are brickmasons, blockmasons, stonemasons, cement masons, concrete finishers. segmental pavers, drywall installers, ceiling tile installers, and tapers.

Plastering one of the oldest crafts in the building trades is enjoying resurgence in popularity because of the introduction of newer, less costly materials and techniques. Plasterers apply plaster to interior walls and ceilings to form fire-resistant and relatively soundproof surfaces. They also apply plaster veneer over drywall to create smooth or textured abrasion-resistant finishes. In addition, plasterers install prefabricated exterior insulation systems over existing walls for good insulation and interesting architectural effects and to cast ornamental designs in plaster.

Present-day stucco masons or plasterers apply durable plasters, such as polymer-based acrylic finishes and stucco to exterior surfaces. But they should not be confused with drywall installers, ceiling tile installers, and tapers, who use drywall instead of plaster when they install interior walls and ceilings.

Plasterers can plaster either solid surfaces, such as concrete block, or supportive wire mesh called lath. When plasterers work with interior surfaces, they first apply a coat of plaster that provides a base, which is followed by a second coat, made of a lime-based plaster. When plastering over metal lath, they apply a preparatory, or *scratch* coat, with a trowel. They spread this rich plaster mixture into and over the metal lath. Before the plaster sets, plasterers scratch its surface with a rake-like tool to produce ridges, so that the subsequent brown coat will bond tightly.

Plasterers spray or trowel this mixture onto the surface, then finish it by smoothing it to an even surface. For the finish coat, they prepare a mixture of lime, plaster of Paris, and water, which they quickly apply this to the brown coat using a *hawk*, a light, metal plate with a handle. This mixture sets very quickly, produces a very smooth, durable finish.

Plasterers can also work with a material that can be finished in a single coat. This *thin-coat* veneer plaster is made of lime and plaster of Paris and is mixed with water at the job site. It provides a smooth, durable, abrasion-resistant finish on interior masonry surfaces, special gypsum baseboard, or drywall prepared with a bonding agent.

Plasterers can create decorative surfaces as well by pressing a brush or trowel firmly against a wet plaster surface and using a circular hand motion to create decorative swirls. For exterior work, stucco masons can embed marble or gravel chips into the finish coat to achieve a pebble-like, decorative finish. Sometimes, plasterers apply insulation to the exteriors of buildings. Also, they cover the outer wall with rigid foam insulation board and reinforcing mesh, and then trowel in a polymer-based or polymer modified base coat. They may apply an additional coat of this material with a decorative finish.

They sometimes do complex decorative and ornamental work that requires special skill and creativity. For example, they may mold intricate wall and ceiling designs, following an architect's blueprint. They can also pour or spray a special plaster into a mold and allow it to set. Workers then remove the molded plaster and put it in place.

Plastering is physically demanding, requiring considerable standing, bending, lifting, and reaching overhead. The work can be dusty and dirty, soiling shoes and clothing, and can irritate the skin and eyes.

Most plasterers and stucco masons work on new construction sites, particularly where special architectural and lighting effects are required. Some repair and renovate older buildings..

Most plasterers and stucco masons work for independent contractors, but about one out of every 10 plasterers and stucco masons is self-employed.

Although most employers recommend apprenticeship as the best way to learn plastering, many people learn the trade by working as helpers for experienced plasterers and stucco masons. Those who learn the trade informally as helpers usually start by carrying materials, setting up scaffolds, and mixing plaster. Later, they learn to apply the scratch, brown, and finish coats.

Apprenticeship programs, sponsored by local joint committees of contractors and unions, generally consist of 2 or 3 years of on-the-job training, in addition to at least 144 hours annually of classroom instruction in drafting, blueprint reading, and mathematics for layout work. In the classroom, apprentices start with a history of the trade and the industry. They also learn about the uses of plaster, estimating materials and costs, and casting ornamental plaster designs. On the job, they learn about lath bases, plaster mixes, methods of plastering, blueprint reading, and safety. They also learn how to use various tools, such as hand and powered trowels, floats, brushes, straightedges, power tools, plaster-mixing machines, and piston-type pumps. Some apprenticeship programs allow individuals to obtain training in related occupations, such as cement masonry and bricklaying.

Applicants for apprentice or helper jobs normally must be at least 18 years old, in good physical condition, and have good manual dexterity, and have a high school education is preferred. Courses in general mathematics, mechanical drawing, and shop provide a useful background. With additional training and experience, plasterers and stucco masons may advance to positions as supervisors, superintendents, or estimators for plastering contractors. Many become self-employed contractors and some become building inspectors.

Job opportunities for plasterers and stucco masons are expected to be good through 2012. Many potential workers may choose not to enter this occupation because they prefer work that is less strenuous and has more comfortable working conditions. The best employment opportunities continue to be in Florida, California, and the Southwest, where exterior plaster and decorative finishes are expected to remain popular. Employment of plasterers and stucco masons is expected to grow about as fast as the average for all occupations through the year 2012. Jobs will become available as plasterers and stucco masons transfer to other occupations or leave the labor force.

In past years, employment of interior plasterers declined as more builders switched to drywall construction. This decline has halted, however, and employment of plasterers is expected to continue growing as a result of the appreciation for the durability and attractiveness that troweled finishes provide. Thin-coat plastering, or veneering, in particular is gaining wide acceptance as more builders recognize its ease of application, durability, quality of finish, and sound-proofing and fire-retarding qualities.

Prefabricated wall systems and new polymer-based or polymer-modified acrylic exterior insulating finishes are also gaining popularity, particularly in the South and Southwest regions of the country. This is not only because of their durability, attractiveness, and insulating properties, but also because of their relatively low cost. In addition, plasterers will be needed to renovate plasterwork in old structures and to create special architectural effects, such as curved surfaces, which are not practical with drywall materials.

Most plasterers and stucco masons work in construction, where prospects fluctuate from year to year due to changing economic conditions. Bad weather affects plastering less than other construction trades because most work is indoors. On exterior surfacing jobs, however, plasterers and stucco masons may lose time because plastering materials cannot be applied under wet or freezing conditions.

In 2002, median hourly earnings of plasterers and stucco masons were $15.91. The middle 50 percent earned between $12.33 and $20.67. The lowest 10 percent earned less than $9.94, and the top 10 percent earned more than $26.81. The median hourly earnings in the largest industries employing plasterers and stucco masons in 2002 were $15.99 in building finishing contractors, and $14.94 in foundation, structure, and building exterior contractors. Apprentice wage rates start at about half the rate paid to experienced plasterers and stucco masons. Annual earnings for plasterers and stucco masons and apprentices can be less than the hourly rate would indicate, because poor weather and periodic declines in construction activity can limit work hours.

Source: Richardson Engineering Services, Process Plant Construction Estimating Standards - 2002.

ANCIENT PLASTERING

PLASTERING DURING THE ROMAN PERIOD

Vitruvius, a Roman engineer of the First Century, provided the following recommendations and specifications for plastering Roman buildings:

"First I shall begin with the concrete flooring, which is the most important of the polished finishings, observing that great pains and the utmost precaution must be taken to ensure its durability. If this concrete flooring is to be laid level with the ground, let the soil be tested to see whether it is everywhere solid, and if it is, level it off and upon it lay the broken stone with its bedding.

But if the floor is either wholly or partly filling, it should be rammed down hard with great care. In case a wooden framework is used, however, we must see that no wall which does not reach up to the top of the house is constructed under the floor. Any wall which is there should preferably fall short, so as to leave the wooden planking above it an unsupported span. If a wall comes up solid, the unyielding nature of its solid structure must, when the joists begin to dry, or to sag and settle, lead to cracks in the floor on the right and left along the line of wall.

Figure A-2
Stucco-finished masonry walls
Ancient Pompeii, Italy

2. We must also be careful that no common oak gets in with the winter oak boards, for as soon as common oak boards get damp, they warp and cause cracks in floors. But if there is no winter oak, and necessity drives, for lack of this, it seems advisable to use common oak boards cut pretty thin; for the less thick they are, the more easily they can be held in place by being nailed on. Then, at the ends of every joist, nail on two boards so that they shall not be able to warp and stick up at the edges. As for Turkey oak or beech or ash, none of them can last to a great age. When the wooden planking is finished, cover it with fern, if there is any, otherwise with straw, to protect the wood from being hurt by the lime.

3. Then, upon this lay the bedding, composed of stones not smaller than can fill the hand. After the bedding is laid, mix the broken stone in the proportions, if it is new, of three parts to one of lime; if it is old material used again, five parts may answer to two in the mixture. Next, lay the mixture of broken stone, bring on your gangs, and beat it again and again with wooden beetles into a solid mass, and let it be not less than three quarters of a foot in thickness when the beating is finished. On this lay the nucleus, consisting of pounded tile mixed with lime in the proportions of three parts to one, and forming a layer not less than six digits thick. On top of the nucleus, the floor, whether made of cut slips or of cubes, should be well and truly laid by rule and level.

4. After it is laid and set at the proper inclination, let it be rubbed down so that, if it consists of cut slips, the lozenges, or triangles, or squares, or hexagons may not stick up at different levels, but be all jointed together on the same plane with one another; if it is laid in cubes, so that all the edges may be level; for the rubbing down will not be properly finished unless all the edges are on the same level plane. The herring-bone pattern, made of Tiber burnt brick, must also be carefully finished, so as to be without gaps or ridges sticking up, but all flat and rubbed down to rule. When the rubbing down is completely finished by means of the smoothing and polishing processes, sift powdered marble on top, and lay on a coating of lime and sand.

5. In the open air, specially adapted kinds of floors must be made, because their framework, swelling with dampness, or shrinking from dryness, or sagging and settling, injures the floors by these changes; besides, the frost and rime will not let them go unhurt. Hence, if necessity drives, we must proceed as follows

In order to make them as free from defects as possible. After finishing the plank flooring, lay a second plank flooring over it at right angles, and nail it down so as to give double protection to the framework.

Then, mix with new broken stone one third the quantity of pounded tile, and let lime be added to the mixture in the mortar trough in the proportion of two parts to five.

6. Having made the bedding, lay on this mixture of broken stone, and let it be not less than a foot thick when the beating is finished. Then, after laying the nucleus, as above described, construct the floor of large cubes cut about two digits each way, and let it have an inclination of two digits for every ten feet. If it is well put together and properly rubbed down, it will be free from all flaws. In order that the mortar in the joints may not suffer from frosts, drench it with oil-dregs every year before winter begins. Thus treated, it will not let the hoarfrost enter it.

7. If, however, it seems needful to use still greater care, lay two-foot tiles, jointed together in a bed of mortar, over the broken stone, with little channels of one finger's breadth cut in the faces of all the joints. Connect these channels and fill them with a mixture of lime and oil; then, rub the joints hard and make them compact.

Thus, the lime sticking in the channels will harden and solidify into a mass, and so prevent water or anything else from penetrating through the joints. After this layer is finished, spread the nucleus upon it, and work it down by beating it with rods. Upon this lay the floor, at the inclination above described, either of large cubes or burnt brick in herring-bone pattern, and floors thus constructed will not soon be spoiled.

Source: Quoted in Norris Hickey Morgan, The Ten Books on Architecture.

THE SLAKING OF LIME FOR STUCCO

1. Leaving the subject of floors, we must next treat of stucco work. This will be all right if the best lime, taken in lumps, is slaked a good while before it is to be used, so that if any lump has not been burned long enough in the kiln, it will be forced to throw off its heat during the long course of slaking in the water, and will thus be thoroughly burned to the same consistency.

When it is taken not thoroughly slaked but fresh, it has little crude bits concealed in it, and so, when applied, it blisters. When such bits complete their slaking after they are on the building, they break up and spoil the smooth polish of the stucco.

But when the proper attention has been paid to the slaking, and greater pains have thus been employed in the preparation for the work, take a hoe, and apply it to the slaked lime in the mortar bed just as you hew wood. If it sticks to the hoe in bits, the lime is not yet tempered; and when the iron is drawn out dry and clean, it will show that the lime is weak and thirsty; but when the lime is rich and properly slaked, it will stick to the tool like glue, proving that it is completely tempered. Then get the scaffolding ready, and proceed to construct the vaultings in the rooms, unless they are to be decorated with flat coffered ceilings.

ANCIENT BUILDING MATERIALS

Thousands of years ago, builders used essentially the same materials as we use today. Of course, without the aid of electric- or gasoline-powered equipment, they had to manufacture and assemble the materials by hand. But just like the modern contractors, they worked with wood and stone and cement. The only significant modern structural material not available to them was prefabricated steel beams, yet they regularly used iron fittings to connect timber and masonry.

Geologists tell us the Holy Land was once at the bottom of a sea. The skeletons and shells of microscopic sea animals sank to the sea floor and were compressed, over the millennia, by the collecting sediment, forming white limestone. When the earth's own turbulence forced this sea floor up above sea level, it formed mountains of limestone and dolomite. As wind, water, and volcanic action eroded this material, it formed various kinds of soils and exposed rock.

These processes, in a way, stocked a "warehouse" full of materials for biblical master builders. We have seen how Solomon had limestone quarried for the Temple, but they also used the sand and clay. Other minerals and metals were mined and developed for use in construction, such as *gypsum*, *pozzolana*, and iron. Of course, trees were also cut and milled into usable lumber.

Master builders dug sand from sand pits, washing it to remove dirt and salt, and then use it in mortar and plaster (beach sand was much too rounded to cohere well). With charcoal fires, they would heat the ores of *lime, gypsum, pozzolana,* and metals, to convert them to construction material.

Clay made great bricks by just adding water. Throughout the Far and Middle Eastern world, workers dug up clay mud, add straw fibers, to increase tensile strength, trampled it to the proper consistency, shaped it with wooden molds; and dried these bricks in the sun. Baked or fired brick, though long known, did not come into general use until the late Roman Republic.

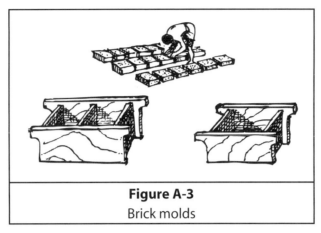

Figure A-3
Brick molds

According to *Vitruvius*, a Roman engineer of the First Century, mud, sun-dried brick had been declared illegal for house walls in Rome, creating a need for kiln-fired clay bricks. Roman bricks came in many shapes and sizes, but the Romans seemed to prefer long, wide bricks that were only an inch and a half thick. These were less likely to warp or crack than thicker bricks.

Figure A-4
Ancient brick-making

STONE MASONRY

Throughout the ancient world, builders used the stones available to them. Construction materials included limestone, marble, sandstone, basalt, granite, and gneiss, among others. Limestone was a nearly perfect material for construction in the Mediterranean World. Plentiful in that area, it was soft and easy to quarry, but it hardened once it became exposed to the environment. To quarry the stone, masons drilled several holes into the rock in a straight line and drove wooden pegs into the holes. When soaked with water, these pegs expanded, exerting pressure on the stone and splitting it in a rather straight line. Using chisels and adzes, masons then shaped the rough faces of the stone. With square plumbs, levels, and measuring strings, they ensured the straightness and angles of the stone block, which would then be sanded with a loaf-like rubbing stone. Other stones were quarried in similar ways.

The stones used in any given locale depended on the geology of that area and the ability to quarry it. We have seen how the early Egyptians used sandstone for the bulk of the pyramids, because that was close at hand. The Babylonians, in the fertile plains of Mesopotamia, were more dependent on clay brick.

The Romans first used *tufa*, a soft tan and brown volcanic rock, in stone-masonry construction. Later they learned to use the harder *lapis gabinus* and *lapis albanus* (now called *sperona* and *peperino*), formed by the action of water on a mixture of volcanic ash, gravel, and sand. Still later, they used a hard, attractive limestone known as *travertine (lapis tiburtinus)*. But unlike the volcanic rocks, it was not fire resistant; under heat, it crumbled into powder.

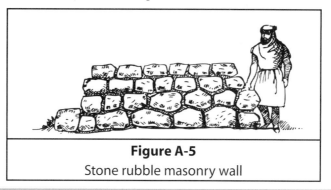

Figure A-5
Stone rubble masonry wall

GYPSUM

As early as 3000 B.C., the Egyptians lined the inside of their pyramids with gypsum plaster. The Greeks also used a gypsum coating in their temples, and the Romans used it in their houses.

In Greek the word *gypsos* means *plaster*, a light-density gray-white rock. It can be ground to a powder and made into *plaster of Paris*. Chemically, the rock is known as dihydrous calcium sulfate ($CaSO_4 \cdot 2H_2O$). Heated in pits, gypsum gives up three-quarters of H_2O as water vapor. As a result of this *calcining*, it can be easily crushed into a talc-like powder. When water is added again and it is mixed, it returns to its original rocklike hardness.

MORTAR

The ancient people made mortar by mixing lime, sand, ashes, and water into a plaster. This was watertight enough to coat cisterns and reservoirs or for use as a finish coating for clay masonry walls and roof decks.

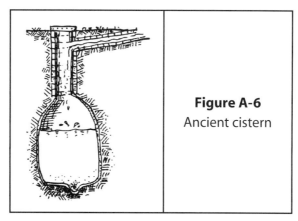

Figure A-6
Ancient cistern

The Romans also used lime mortar between bricks and stone blocks on occasion. But in their public buildings, they generally relied upon iron clamps and the close fit of the neatly trimmed stones to hold the structures together.

CONCRETE

Near Vesuvius and elsewhere in Italy were deposits of a sandy volcanic ash. When added to lime mortar, this made a cement that would harden even under water. They named this material *pulvis puleoli*, after Puteoli, an area that contained huge beds of it. By mixing this cement with sand and gravel, the Romans created a sturdy concrete.

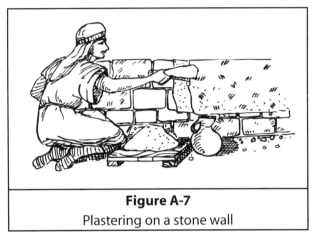

Figure A-7
Plastering on a stone wall

But initially the Romans apparently did not realize the possibilities of this new concrete, because they used it only in small amounts. Builders continued to apply mortars and plasters for centuries, until the new waterproof concrete began to catch on.

Vitruvius described in detail the importance of the quality of sand used in mortar and concrete:

"In walls of masonry the first question must be with regard to the sand, in order that it may be fit to mix into mortar and have no dirt in it. The kinds of pit sand are these: black, grey, red, and carbuncular.... The best… crackles when rubbed in the hand, while that which has much dirt in it will - not be sharp enough….

But if there are no sand pits from which it can be dug, then we must sift it out of river beds or from gravel or even from the sea beach. This kind, however, has these defects when used in masonry: it dries slowly… and such a wall cannot carry vaulting….

But pit sand used in masonry dries quickly, the stucco coating is permanent, and the walls will support vaulting. I am speaking of sand fresh from the sand pits. For if it lies unused too long after being taken out, it is disintegrated by exposure… and becomes earthy…. Fresh pit sand, however, in spite of all its excellence in concrete structures, is not equally useful in stucco, the richness of which, when the lime and straw are mixed with such sand, will cause it to crack as it dries on account of the great strength of the mixture. But river sand, though useless in "signinum" on account of its thinness, becomes perfectly solid in stucco when thoroughly worked by means of polishing instruments."

Source for Figures A-3 through A-7:
Max Schwartz, "Exploring Buildings, Machines, and Weaponry in
Biblical Times, Eerdman Publishing 1990.

APPENDIX B

Cement Manufacture

PORTLAND CEMENT MANUFACTURE

The industrial uses of limestone and cements have provided important undertakings for chemists and engineers since the early years when lime mortars and natural cements were first introduced. In modern times, one needs only mention reinforced concrete walls and girders, tunnels, dams, and roads to realize the dependence of present-day civilization upon these products. The convenience, cheapness, adaptability, strength, and durability of both lime mortar and cement products have been the foundation of these tremendous applications. The once crude and intermittent type of lime and cement production has been perfected to a point where countless millions of tons were produced in the United States.

CEMENT

In spite of our modern concrete roads and buildings everywhere around us, it is difficult to realize the tremendous growth of the cement industry during the past century. Man had early discovered certain natural rocks, which through simple calcination, produced a product that hardened on the addition of water. Yet the real advance did not take place until chemical analysis and engineering laid the basis for the modern efficient plants working under closely controlled conditions on a variety of raw materials.

In 1824, an Englishman, Joseph Aspdin, patented an artificial cement, which he called Portland because concrete made from it resembled a famous building stone obtained from the Isle of Portland near England. Very shortly, off the coast of Harwick, hundreds of boats were engaged in dredging for "cement stones." These were *argillaceous* limestone, which upon being burned, would harden when water was added and not fall to a powder as would ordinary limestone.

Cement so made, irrespective of brand, is known as *Portland cement* to distinguish it from *natural* or *pozzolana* and other cements.

A century ago, concrete was little used in this country because the manufacture of Portland cement was a complicated and expensive process. Thanks to the invention of laborsaving machinery, cement is now low in cost and is applied everywhere in the construction of homes, public buildings, roads, industrial plants, dams, bridges, and in many other places.

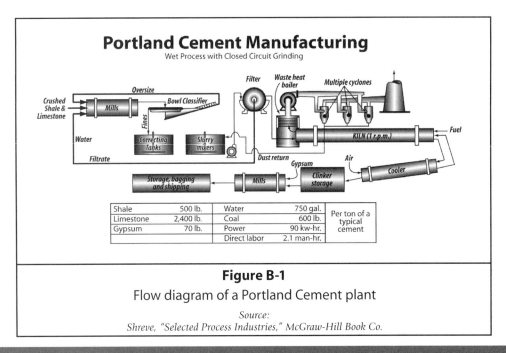

Figure B-1
Flow diagram of a Portland Cement plant

Source:
Shreve, *"Selected Process Industries,"* McGraw-Hill Book Co.

TYPES OF CEMENT

The various types of cements are:

1. *Pozzolana cement:* Since the beginning of the Christian era, the Italians have successfully employed pozzolana cement, made by grinding two to four parts of a *pozzolana* with one part of hydrated lime. A *pozzolana* is a material, which is not cementitious in itself, but which becomes so upon admixture with lime. The natural pozzolans are volcanic tuffs, and the artificial ones are burnt clays and shales.

2. *Portland cement* is the product obtained by finely pulverizing clinker made by calcining to incipient fusion an intimate and properly proportioned mixture of *argillaceous* and *calcareous* materials with no additions subsequent to calcination excepting water and calcined or uncalcined gypsum. This is a flour-like powder, which varies in color from a greenish gray to a brownish gray.

Portland cement, may be subdivided into five classes:

- *Regular Portland cements* are the usual products required to meet no exacting or special demands. There is a white Portland cement sold, which contains less iron.
- *Masonry cements* are mixtures of cement with hydrated lime, crushed limestone, diatomaceous earth, with or without small addition of *calcium stearate*, petroleum, or highly *colloidal* clays. These are easily workable because of increased plasticity.
- Plastic cement (ASTM Designation 1328) is a product manufactured expressly for the plaster industry and is most commonly available in the Southwest and along the West Coast. It consists of a mixture of Portland cement and blended hydraulic cement and plasticizing materials (such as limestone, hydrated or hydraulic lime), together with other materials to enhance one or more properties such as setting time, workability, water retention and durability. Typically, when plastic cement is used, no lime or other plasticizer may be added to the plaster at the time of mixing.

- *High-early-strength (H.E.S.) cements* are made from a raw material with a high lime to silica ratio. They are frequently burned twice and is very finely ground. They contain a higher proportion of *tricalcium silicate*, C_3S, than regular Portland cements and hence harden much more quickly and with greater evolution of heat. Roads constructed from H.E.S. cement can be put into service more quickly than if regular cement had been employed.
- Low-heat Portland cements contain a higher percentage of *tetracalcium aluminoferrite*, C_4AF, and *dicalcium silicate*, C_2S, and hence set with the evolution of much less heat. Also, *tricalcium silicate*, C_3S, and the *tricalcium aluminate*, C_3A, are lower. Actually the heat evolved should not exceed 60 and 70 cal. per gram after 7 and 28 days, respectively, and is 15 to 35 per cent less than the heat of hydration of regular or H.E.S. cements.

There are also available *chemical-* or *sulfate-resisting Portland cements,* which by their composition or processing resist chemicals better than the other three types. These cements are higher in *tetracalcium aluminoferrite*, C_2F, and lower in *tricalcium aluminate*, C_3A, than the regular cements. Additions during grinding of small percentages of *calcium stearate* or *sodium silicate* are somewhat effective. There is, however, no such thing as an acid-proof Portland cement.

Plastic cement is most commonly found in the Southwestern and Pacific States. It is a hydraulic cement that is used primarily in Portland cement-based plastering construction (such as stucco). Plastic cement is a mixture of Portland and blended hydraulic cement, as well as plasticizing materials such as limestone and hydrated or hydraulic lime, along with other additions whose purpose is to control such properties as setting time, workability, water retention and durability. ASTM C 1328 defines sets out separate requirements for a Type M and a Type S plastic cement; UBC 25-1 does not differentiate between the two types of plastic cement, and instead delineates a single requirement that corresponds to the ASTM C 1328 Type M plastic cement standard. When plastic cement is used, no lime or other plasticizer should be added at the time of mixing. In the Southeastern U.S., plastic cement is known as "stucco cement" and typically is formulated to meet ASTM C 1328 Type S requirements, to produce a product that is more appropriate for the aggregates, environmental factors and construction practices that are typical of that region.

High-alumina cement is manufactured by fusing a mixture of limestone and *bauxite,* the latter usually containing *iron oxide, silica, magnesia,* and other impurities. It is characterized by a very rapid rate of development of strength and superior resistance to seawater and sulfate-bearing water.

Special or acid-resisting cements are on the market and find use in erecting acid washing towers and for mortar in laying floors or walls to resist acids. A high-silica water glass mixed into a dough with silica sand or coarse asbestos powder has found favor. This is frequently "set" with a strong sulfuric acid wash. Likewise molten sulfur containing some filler such as asbestos or sand is poured into brick joints.

Sources: Shreve, "Selected Process Industries" McGraw-Hill Book Co.; "PCA Manual" Portland Cement Association.

LIME MANUFACTURE

Although lime has been employed as a building material for centuries, very little exact information about its properties has been available until recently. The manufacture of lime and its use can be traced back throughout Roman, Greek, and Egyptian civilizations. The first definite written information concerning lime was handed down from the Romans.

In his book, *De Architectura,* Marcus Pollio, a celebrated engineer and architect who lived during the reign of Augustus (27 B.C. to A.D. 14) dealt quite thoroughly with the use of lime for mortar involved in the construction of harbor works, pavements, and buildings.

In the early days of the young American colony, the crude burning of limestone was one of the initial manufacturing processes engaged in by the settlers, using *dug-out* kilns built of ordinary brick or masonry in the side of a hill, with a coal or wood fire at the bottom and a firing time of 72 hours.

These kilns can still be seen in many of the older sections of the country. It was not until recent years that, under the influence of scientific cooperative research, the manufacture of lime has developed into a large industry under exact technical control, with the resulting uniformity of products.

USES AND ECONOMICS

Lime itself may be used for medicinal purposes, insecticides, plant and animal food, gas absorption, precipitation, dehydration, and *causticizing*. It is also employed as a reagent in the sulfite process for papermaking, dehairing hides, recovering by-product ammonia, manufacturing of high-grade steel and cement, water softening, manufacturing of soap, rubber, varnish, and refractories, and the making of sand-lime brick.

Lime is indispensable for mortar and plaster use and serves as a basic raw material for calcium salts and for improving the quality of certain soils.

Lime is sold as a high-calcium quicklime containing not less than 90 per cent of calcium oxide and from 0 to 5 per cent of *magnesia* with small percentages of *calcium carbonate, silica, alumina,* and *ferric oxide* present as impurities. The suitability of lime for any particular use depends on its composition and physical properties, all of which can be controlled by the selection of the limestone and the detail of the manufacturing process.

Lime must be finely ground before use. Depending on composition., there are several distinct types of limes Hydraulic limes are obtained from the burning of limestone containing clay, and the nature of the product obtained after contact with water varies from a putty to a set cement.

High-calcium-content limes harden only by absorption of carbon dioxide from the air, which is a slow process. Hydraulic limes also harden slowly but they can be used under water. For chemical purposes high-calcium lime is required except for the sulfite paper process where a magnesium lime works better.

Although in many sections of the country the high-calcium lime is preferred by the building industry for the manufacture of its mortar or its lime plaster, there are places where limestone containing some magnesium is burned or where even a dolomitic stone is calcined.

Typical compositions of these products, called *magnesium limes* or *dolomitic limes,* find favor in the hands of some plasterers who claim they work better under the trowel. In the metallurgical field, "refractory lime" as dead-burned dolomite or as raw dolomite is employed as a refractory patching material in open-hearth furnaces, being applied between heats to repair scored and washed spots in the bottom of the furnace.

Hydrated lime is finding increased favor in the chemical and building trades over the less stable quicklime, despite its increased weight. The quicklime is almost invariably slaked or hydrated before use.

LIME MANUFACTURE

Lime has always been a cheap commodity because limestone deposits are readily available in so many sections of the United States and hence permit its burning near centers of consumption.

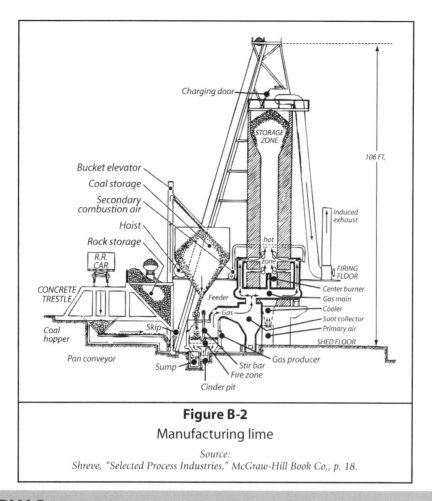

Figure B-2
Manufacturing lime

Source:
Shreve, "Selected Process Industries," McGraw-Hill Book Co., p. 18.

RAW MATERIALS

The *carbonates of calcium* or *magnesium* are obtained from naturally occurring deposits of limestone, marble, chalk, or dolomite. For chemical usage, a rather pure limestone is preferred as a starting material because of the high-calcium lime that results. The quarries furnish a rock that contains as impurities low percentages of silica, clay, or iron. Such impurities are important because the lime may react with the silica and alumina to give *calcium silicates* or *calcium alumino-silicates* which possess not undesirable hydraulic properties.

The lumps sometimes found in "overburned" or "dead-burned" lime prevent "popping out," the MgO must be completely hydrated such as by steam hydration at 160° C. and under such pressure as 60 lb. per sq. in. tabulates the results from grinding and air-separating hydrated lime from much of the impurities originally present in the quicklime wherein the purity is raised from 89 to 98 per cent, the tailing going to agricultural applications result from changes in the calcium oxide itself as well as from certain impurities acted upon by excess heat, recognized as masses of relatively inert, semi-vitrified material. On the other hand, it often happens that rather pure limestone is calcined insufficiently, and lumps of calcium carbonate are left in the lime. This lime is spoken of as "underburned" lime.

APPENDIX C

Governing Codes and Specifications

CODES, STANDARDS AND GUIDES

Construction specifications setting forth the minimum requirements for plastering work must be closely followed, whether stated in industry-wide standards and guide specifications or in local building codes and ordinances. These written specifications establish acceptance requirements for the quality of materials and workmanship.

Compliance with construction specifications protects both the customer and contractor. All members of the design and construct team should be aware that design, materials, and construction affect the performance of each segment of a structure as well as the entire structure.

Though applicable codes vary throughout North America, they are always mandatory and should be strictly followed. In the past in the United States, local jurisdictions adopted one of three model building codes, the BOCA National Building Code (BOCA), the Standard Building Code (SBC), or the Uniform Building Code (UBC). In Canada, specifying agencies typically follow the National Building Code, which contains stucco information in Section 9.28.

During the 1990s in the United States, the sponsoring organizations that produced the BOCA, SBC, and UBC formed the International Codes Council (ICC) and developed the International Building Code (IBC) for commercial construction and the International Residential Code (IRC) for dwellings. Working from the existing three model codes, building officials established a unified code for the country and beyond.

Even with the new 2000 I-Codes barely in effect, a competing model building code from the National Fire Protection Agency, designated NFPA 5000, was completed in 2002. However, the Portland cement plaster provisions of BOCA, SBC, IBC, IRC and NFPA 5000 reference ASTM C 926, the Standard Specification for Application of Portland Cement-Based Plaster, and ASTM C 1063, the Standard Specification for Installation of Lathing and Furring for Portland Cement-Based Plaster.

The UBC does not reference these two ASTM Standards. UBC provisions for Portland cement plaster are contained in Chapter 25 of that model code.

Project specifications typically contain additional references to national standards and industry and manufacturers' guides and recommendations. Current standards pertaining to Portland cement plaster and plastering, approved by ASTM and the American Concrete Institute (ACI), should be reviewed and understood before commencing any phase of plastering. Manufacturers' guides and recommendations address plastering while concentrating on a specific material. As guides and recommendations, their contents are informative but not mandatory, unless required by the project specification.

C-35 PLASTERING CONTRACTOR DEFINED

A plastering contractor is a *specialty contractor* whose principal business is the execution of contracts, usually subcontracts, requiring a knowledge of the art and science of coating surfaces with the mixture of:

- Sand, gypsum plaster, quicklime, or hydrated lime and water, or sand and cement and water, or
- Combination of such other materials as to create a permanent surface coating.

Such coatings are usually applied with a plasterer's trowel over any surface, which offers a mechanical key for the support of such coating, or such coating will adhere by suction, or the affixation of lath or any other material or product prepared or manufactured to provide a base for such coating.

C-35 PLASTERING CONTRACTOR

(a) A lathing and plastering contractor coats surfaces with a mixture of sand, gypsum plaster, quick-lime or hydrated lime and water, or sand and cement and water, or a combination of such other materials that create a permanent surface coating, including coatings for the purpose of soundproofing and fireproofing. These coatings are applied with a plasterer's trowel or sprayed over any surface which offers a mechanical means for the support of such coating, and will adhere by suction. This contractor also installs lath (including metal studs) or any other material prepared or manufactured to provide a base or bond for such coating.

(b) A lathing and plastering contractor also applies and affixes wood and metal lath, or any other material prepared or manufactured to provide key or suction bases for the support of plaster coatings. This classification includes the channel work and metal studs for the support of metal or any other lathing material and for solid plaster partitions.

(c) Effective January 1, 1998, or as soon thereafter as administratively feasible, all C-26 licensees will be merged into the C-35 Lathing and Plastering classification. On and after January 1, 1998, no application for the C-26 classification will be accepted and no new C-26 Lathing licenses will be issued.

SECTION 2506A - EXTERIOR LATH

2506A.1   General

Exterior surfaces are weather-exposed surfaces as defined in Section 224. For eave overhangs required to be fire resistive, see Section 705.

2506A.2   Corrosion Resistance

All lath and lath attachments shall be of corrosion-resistant material. See Section 2501A.4.

2506A.3   Backing

Backing or a lath shall provide sufficient rigidity to permit plaster application. Where lath on vertical surfaces extends between rafters or other similar projecting members, solid backing shall be installed to provide support for lath and attachments.

Backing is not required under metal lath or paperbacked wire fabric lath.

2506A.4 Weather-resistive Barriers

Weather-resistive barriers shall be installed as required in Section 1402A.l and, when applied over wood base sheathing, shall include two layers of Grade D paper.

2506A.5 Application of Metal Plaster Bases

The application of metal lath or wire fabric lath shall be as specified in Section 2505A.3 and they shall be furred out from vertical supports or backing not less than 1/4 inch (6.4 mm) except as set forth in Table 25A-B, Footnote 2. In DSA (Calif. State Division of State Architect) involved construction, where exterior lath is attached to horizontal wood supports, either of the following attachments shall be used in addition to the methods of attachment set forth in Table 25A -C:

1. Secure lath to alternate supports with ties consisting of a double strand of No. 18 W&M gauge galvanized annealed wire at one edge of each sheet of lath.

Wire ties shall be installed not less than 3 inches (76 mm) back from the edge of each sheet and shall be looped around stripping, or attached to an 8d common wire nail driven into each side of the joist 2 inches (51 mm) above the bottom of the joist or to each end of a 16d common wire nail driven horizontally through the joist 2 inches (51 mm) above the bottom of the joist and the ends of the wire secured together with three twists of the wire.

2. Secure lath to each support with 1/2-inch-wide (12.7 mm), 1½-inch-long (38 mm) No. 9 W&M gauge ring shank, hook staple placed around a 10d common nail laid flat under the surface of the lath not more than 3 inches (76 mm) from edge of each sheet. Such staples may be placed over ribs of 3/8-inch (9.5 mm) rib lath or over back wire of welded wire fabric or other approved lath, omitting the 10d nails.

Where no external comer reinforcement is used, lath shall be furred out and carried around comers at least one support on frame construction.

A weep screed shall be provided at the foundation plate line on all exterior stud walls. The screed shall be of a type which will allow trapped water to drain to the exterior of the building. The weather-resistant barrier and exterior lath shall cover and terminate on the attachment flange of the screed.

A minimum 0.019-inch (0.48 mm) (No. 26 galvanized sheet gauge) corrosion-resistant weep screed with a minimum vertical attachment flange of 3-1/2 inches (89 mm) shall be provided at or below the foundation plate line on all exterior stud walls. The screed shall be placed a minimum of 4 inches (102 mm) above the earth or 2 inches (51 mm) above paved areas and shall be of a type that will allow trapped water to drain to the exterior of the building. The weather-resistive barrier shall lap the attachment flange, and the exterior lath shall cover and terminate on the attachment flange of the screed.

SECTION 2508A - EXTERIOR PLASTER

General

Plastering with cement plaster shall not be less than three coats when applied over metal lath or wire fabric lath and shall not be less than two coats when applied over masonry, concrete or gypsum backing as specified in Section 2506A.3 of the UBC.

If plaster surface is completely covered by veneer or other facing material, or is completely concealed by another wall, plaster application need be only two coats, provided the total thickness is as set forth in Table 25A-F.

On wood-frame or metal stud construction with an on-grade concrete floor slab system, exterior plaster shall be applied in such a manner as to cover, but not extend below, lath and paper. See Section 2506A.5 for the application of paper and lath, and flashing or weep screeds.

Only approved plasticity agents and approved amounts thereof may be added to Portland cement. When plastic cement is used, no additional lime or plasticizers shall be added. Hydrated lime or the equivalent amount of lime putty used as a plasticizer may be added to cement plaster or cement and lime plaster in an amount not to exceed that set forth in Table 25A-F.

Gypsum plaster shall not be used on exterior surfaces.

2508A.2   Base Coat Proportions

The proportion of aggregate to cementitious materials shall be as set forth in Table 25A-F.

2508A.3 Base Coat Application.

The first coat shall be applied with sufficient material and pressure to fill solidly all openings in the lath. The surface shall be scored horizontally sufficiently rough to provide adequate bond to receive the second coat.

The second coat shall be brought out to proper thickness, rodded and floated sufficiently rough to provide adequate bond for the finish coat. The second coat shall have no variation greater than 1/4 inch (6.4 mm) in any direction under a 5-foot (1524 mm) straight edge.

2508A.4   Environmental Conditions

Portland cement-based plaster shall not be applied to frozen base or those bases containing frost. Plaster mixes shall not contain frozen ingredients. Plaster coats shall be protected from freezing for a period of not less than 24 hours after set has occurred.

2508A.5   Curing and Interval

First and second coats of plaster shall be applied and moist cured as set forth in Table 25A-F.

When applied over gypsum backing as specified in Section 2506A.3 or directly to unit masonry surfaces, the second coat may be applied as soon as the first coat has attained sufficient hardness.

2508A.6   Alternate Method of Application

As an alternate method of application, the second coat may be applied as soon as the first coat has attained sufficient rigidity to receive the second coat. When using this method of application, calcium aluminate cement up to 15 percent of the weight of the Portland cement may be added to the mix.

Curing of the first coat may be omitted and the second coat shall be cured as set forth in Table 25A-F.

2508A.7   Finish Coats

Finish coats shall be proportioned and mixed in an approved manner and in accordance with Table 25A-F.

Cement plaster finish coats shall be applied over base coats that have been in place for the time periods set forth in Table 25A-F.

The third or finish coat shall be applied .with sufficient material and pressure to bond to and to cover the brown coat and shall be of sufficient thickness to conceal the brown coat.

2508A.8 Preparation of Masonry and Concrete

Surfaces shall be clean, free from efflorescence, sufficiently damp and rough to ensure proper bond. If surface is insufficiently rough, approved bonding agents or a Portland cement dash bond coat mixed in proportions of one and one half parts volume of sand to one part volume of Portland cement or plastic cement shall be applied. Approved bonding agents shall conform with the provisions of United States Government Military Specifications MIL-B-19235.

Dash bond coat shall be left undisturbed and shall be moist cured not less than 24 hours. When dash bond is applied, first coat of base coat plaster may be omitted. See Table 25A-D for thickness.

SECTION 2509A - EXPOSED AGGREGATE PLASTER

2509A.1 General

Exposed natural or integrally colored aggregate may be partially embedded in a natural or colored bedding coat of cement plaster or gypsum plaster, subject to the provisions of this section.

2509A.2 Aggregate

The aggregate may be applied manually or mechanically and shall consist of marble chips, pebbles or similar durable, nonreactive materials, moderately hard (three or more on the Mohs scale).

2509A.3 Bedding Coat Proportions

The exterior bedding coat shall be composed of one part Portland cement, one part Type S lime, and a maximum three parts of graded white or natural sand by volume. The interior bedding coat shall be composed of 100 pounds (45.4 kg) neat gypsum plaster and a maximum 200 pounds (90.7 kg) of graded white sand, or exterior or interior may be a factory-prepared bedding coat. The exterior bedding coat shall have a minimum compressive strength of 1,000 pounds per square inch (6894.8 kPa).

2509A.4 Application

The bedding coat may be applied directly over the first (scratch) coat of plaster, provided the ultimate overall thickness is a minimum of 7/8 inch (22.2 mm), including lath.

Over concrete or masonry surfaces, the overall thickness shall be a minimum of 1/2 inch (12.7 mm).

2509A.5 Bases

Exposed aggregate plaster may be applied over concrete, masonry, cement plaster base coats or gypsum plaster base coats.

2509A.6 Preparation of Masonry and Concrete

Masonry and concrete surfaces shall be prepared in accordance with the provisions of Section 2508A.8.

2509A.7 Curing

Cement plaster base coats shall be cured in accordance with Table 25A-F. Cement plaster bedding coat shall retain sufficient moisture for hydration (hardening) for 24 hours minimum or, where necessary, shall be kept damp for 24 hours by light water spraying.

Table C-1 / Cement Plasters[1]

PORTLAND CEMENT PLASTER

Coat	Volume Cement	Maximum Weight (or Volume) Lime per Volume Cement	Maximum Volume Sand per Combined Volumes Cement and Lime[2]	Approximate Minimum Thickness[3] (x 25.4 for mm)	Minimum Period Moist Curing	Minimum Interval between Coats
First	1	20 lbs. (9.07 kg)	4	3/8"[4]	48 hours[5]	48 hours[6]
Second	1	20 lbs. (9.07 kg)	5	1st and 2nd coats total 3/4"	48 hours	7 days[7]
Finish	1	1[8]	3	1st, 2nd and finish coats 7/8"	—	7

PORTLAND CEMENT - LIME PLASTER[9]

Coat	Volume Cement	Maximum Volume Lime per Volume Cement	Maximum Volume Sand per Combined Volumes Cement and Lime[3]	Approximate Minimum Thickness[3] (x 25.4 for mm)	Minimum Period Moist Curing	Minimum Interval between Coats
First	1	1	4	3/8"[4]	48 hours[5]	48 hours[6]
Second	1	1	4½	1st and 2nd coats total 3/4"	48 hours	7 days[7]
Finish	1	1[8]	3	1st, 2nd and finish coats 7/8"	—	7

PLASTIC CEMENT PLASTER[9]

Coat	Volume Cement	Maximum Weight (or Volume) Lime per Volume Cement	Maximum Volume Sand per Volume Cement[2]	Approximate Minimum Thickness[3] (x 25.4 for mm)	Minimum Period Moist Curing	Minimum Interval between Coats
First	1	—	4	3/8"[4]	48 hours[5]	48 hours[6]
Second	1	—	5	1st and 2nd coats total 3/4"	48 hours	7 days[7]
Finish	1	—	3	1st, 2nd and finish coats 7/8"	—	7

[1] Exposed aggregate plaster shall be applied in accordance with Section 2509A. Minimum overall thickness shall be 3/4 inch (19 mm)
[2] When determining the amount of sand in set plaster, a tolerance of 10 percent may be allowed.
[3] See Table 25A-D.
[4] Measured from face of support or backing to crest of scored plaster.
[5] See Section 2507A.3.3.
[6] Twenty-four-hour minimum interval between coats of interior cement plaster. For alternate method of application, see Section 2508A.6.
[7] Finish coat plaster may be applied to interior Portland cement base coats after a 48-hour period.
[8] For finish coat plaster, up to an equal part of dry hydrated lime by weight (or an equivalent volume of lime putty) may be added to Types I, II and III standard Portland cement.
[9] No additions of plasticizing agents shall be made.

Source: California Building Code, 2001.

GUIDE SPECIFICATIONS

Source: Stucco Manufacturers Association, and California Lathing and Plastering Contractors Association, Inc.

10. STUCCO FINISHES

10.1 GENERAL REQUIREMENTS

10.1.1 Packaging

Each manufacturer shall pack manufactured stucco in sealed, multi-wall bags bearing his name, brand, weight and color identification.

10.1.2 Additives

No fire clay, asbestos, or any other material except clean water shall be added to manufactured stucco.

10.2 EXTERIOR STUCCO (See Note 1)

10.2.1 Uses

Over any properly prepared Portland cement base (plaster, concrete or masonry).

10.2.2 Materials

A packaged blend of Portland cement (ASTM C-150), hydrated lime (C-206) and properly graded quality aggregate, with or without color.

10.2.3 Properties

When tested per ASTM C-109-63, exterior stucco shall have a minimum compressive strength of 1200 psi.

10.3 FINISH PLASTER BEDDING COAT (See Note 2)

10.3.1 Uses

As a bedding coat to receive exposed aggregate.

10.3.2 Materials

A packaged blend of Portland cement (ASTM C-150), hydrated lime (C-206) and properly graded quality aggregate, with or without color.

10.3.3 Properties

When tested per ASTM C-109-63, Finish plaster bedding coat shall have a minimum compressive strength of 2000 psi.

10.4 SMOOTH CEMENT FINISH

10.4.1 Uses

Over any properly prepared Portland cement base (plaster, concrete or masonry). In wet rooms and high humidity areas (shower rooms, lavatories, etc.) or in areas subject to extreme abuse (handball courts, etc.).

10.4.2 Materials

A packaged blend of Portland cement (ASTM C-150), hydrated lime (C-206) and properly graded quality aggregate, without color.

10.4.3 Properties

When tested per ASTM C-109-63, smooth cement finish shall have a minimum compressive strength of 2000 psi. This is the most difficult texture to obtain without excessive cracking and uneven shade and color. A glass fiber mesh should be embedded in 1/8-in. of cement adhesive on top of properly floated and cured basecoat plaster.

10.4.4 Curing

Except during damp weather, surface shall be dampened slightly 12 hours after completion and re-dampened at intervals until it hardens.

10.5 PORTLAND CEMENT STUCCO PAINT (See Note 3)

10.5.1 Uses

On porous surfaces of masonry, concrete, stucco, common brick, masonry block and rough plaster as a decorative, protective, and water-repellent coating.

10.5.2 Materials and Properties

Shall conform to Federal Specification TT-P-21.

10.6 ACOUSTIC TYPE FINISH (EXTERIOR) (See Note 4)

10.6.1 Uses

Over any properly prepared Portland cement base, such as plaster, concrete, or masonry where texture of interior acoustic ceilings must be matched, or where such a texture is desired on exterior horizontal surfaces. Specify only thickness required to achieve desired texture.

10.6.2 Materials

A packaged blend of Portland cement (ASTM C-150), hydrated lime (C-206) and vermiculite aggregate (C-35).

NOTE 1. Special Recommendations
  A. For color control add measured amounts of water required to maintain uniform consistency for type of texture specified.
  B. Light pastels are recommended with a float texture. Darker colors may be specified with dash or troweled textures.
  C. Clean mixer thoroughly between color changes.
  D. Curing is necessary under hot, dry or windy conditions. When required, fog lightly the day following application.

NOTE 2. Special Recommendations
  A. For color control add measured amounts of water required to maintain uniform consistency.
  B. Clean mixer thoroughly between color changes.
  C. Curing is necessary under hot, dry or windy conditions. When required, fog lightly the day following application.
  D. Specify penetrating sealer or non-penetrating glaze to enhance colors and protect surface.

NOTE 3. Special Recommendations
  A. Mix material by adding water slowly and stirring until a thick paste is formed. Allow to stand for 10 minutes. Add more water to obtain smooth flowing mixture slightly thicker than milk for hand application, water consistency for spraying. Water proportion must remain constant to produce uniform color.
  B. Hand application requires a large brush. Machine application is by pressure spray equipment, one coat for a color similar to existing surface, two when specifying different color.
  C. Except during damp weather, surface shall be dampened at intervals until it hardens.

NOTE 4. Special Recommendations
  A. Mix in a mechanical mixer until fluffy (at least 10 minutes).

B. Specify two coats applied by plaster machine. Spray second coat to uniform texture when visible moisture has left surface.

C. Curing is necessary except during damp weather. When required, fog slightly the day following application.

NOTE 5. **Special Recommendations**

A. For color control add measured amounts of water required to maintain uniform consistency for type of texture specified.

B. Light pastels are recommended with a smooth finish. Darker colors may be specified with float, dash or troweled textures.

C. Clean mixer thoroughly between color changes.

NOTE 6. **Special Recommendations**

A. Mix in a mechanical mixer until fluffy (at least 10 minutes).

B. Specify two coats applied by plaster machine. Spray second coat to a uniform texture when visible moisture has left surface.

10.10 MARBLE CRETE

Source: California Lathing & Plastering Contractors Association, Inc.

10.10.1 Description

Finish plaster shall consist of exposed natural or integrally colored aggregate, partially embedded in a natural (or) colored bedding coat of Portland cement plaster.

10.10.2 Location

Apply finish plaster in areas where shown on the drawings, or called for in the finish schedule or in these specifications. (See Note 1.)

10.10.3 Materials

Lathing and plastering materials for finish plaster shall be standard lathing and plastering materials. (See Note 2.)

10.10.31 Aggregate

For finish plaster finish shall consist of marble chips or pebbles, and shall be clean and free from harmful amounts of dust and other foreign matter. (See Note 3.) Size of finish plaster aggregate shall conform to the following grading standard:

Chip Size	Passing Screen	Retained on Screen
Number	Inches	Inches
0	1/8	1/16
1	1/4	1/8
2	3/8	1/4
3	1/2	3/8
4	5/8	1/2

NOTE: Chips larger than No. 4 must be applied manually. Use only for random accent. Check with stucco manufacturer for local availability of aggregate desired.

Aggregate shall be blended by sizes in the percentages of graded chips called for in the specified sample.

10.10.32 Liquid Bonding Agent

See 6.10.1. California Reference Specifications. (See Note 4.)

10.10.33 Sealer

Waterproofing shall be a clear penetrating liquid; (or) glaze shall be a clear non-penetrating liquid. Sealer shall be non-staining and shall resist deterioration from weather exposure.

Apply as directed by the manufacturer. (See Notes.)

10.10.4 Bases

Apply finish plaster over (a) concrete, (b) masonry, (c) Portland cement plaster base coat, (d) gypsum plaster basecoat. (See Note 6.)

10.10.41 Concrete & Masonry

Give masonry and poured concrete surfaces which are to receive finish plaster a dash bond coat of Portland cement plaster; or treat with a liquid bonding agent. (See Note 7.)

Apply plaster using one of the following option methods: (See Note 8).

(a) Apply bedding coat over dash bond coat or liquid bonding agent and double back to required thickness.

(b) Apply brown (leveling) coat over dash bond coat or liquid bonding agent and straighten with rod and darby before applying bedding coat.

10.10.42 Basecoat Plaster

Over metal or wire fabric lath apply basecoat plaster in one of the following optional methods:

(a) Apply scratch coat and allow to set. Apply brown (leveling) coat minimum three-eighth inch (⅜ II) ,thick and straighten with rod and darby. Leave rough and allow to set before applying bedding coat.

(b) Apply same as above except that bedding coat may be applied as soon as brown coat is firm enough to support bedding coat without sagging, sliding, or otherwise affecting bond.

(c) Apply scratch coat minimum one-half inch (1/2") thick using double-back method, cover lath completely and straighten with rod and darby. Scratch horizontally and allow to set before applying bedding coat. Minimum overall thickness of scratch and bedding coat shall be one inch (1").

10.10.5 Bedding Coat

Factory prepared bedding coat shall be a Portland cement and lime plaster meeting "Specifications and Standards for Manufactured Stucco Finishes" as published by the Stucco Manufacturers Association, Inc., 15926 Kittridge, Van Nuys, Calif. (See Note 9.)

Job proportioned bedding coat shall be composed of one part Portland cement, one part Type S lime, and maximum three parts of graded white or natural sand by volume. (See Note 10.)

10.10.51 Mixing

Mix manufactured and job proportioned bedding coat with only sufficient water to attain proper consistency for application and embedment of aggregate.

10.10.52 Thickness

Thickness of bedding coat shall be determined by maximum size of aggregate specified and shall conform to the following: (See Note 11).

Bedding Coat Thickness (Minimum)	Aggregate Size (Maximum)	
	Number	Inches
3/8	# 0	1/8
3/8	# 1	1/4
3/8	# 2	3/8
3/8	# 3	1/2
1/2	# 4	5/8
80% of aggregate dimension	Larger than #4	

10.10.6 Application of Finish plaster

Apply bedding coat to proper thickness and straighten to a true, reasonably smooth surface with rod and darby.

Allow bedding coat to take up until it attains the proper consistency to permit application of aggregate.

Apply the aggregate to the bedding coat, starting at the perimeter of a panel area and working towards the center.

Tamp lightly and evenly to assure embedment of the aggregate and to bring surface to an even plane.

10.10.7 Curing

Portland cement finish plaster shall retain sufficient moisture for hydration (hardening) for 24 hours minimum. Where weather conditions require, keep finish plaster damp by spraying lightly.

10.10.8 Samples

Furnish samples in duplicate (triplicate). Samples shall conform to the following:

A. Screed and Casing Thickness   Inch

B. Bonding Agent

C. (1) Bedding Coat Thickness   Inch

    (2) Bedding Coat Color Number   Number, by Mfr.

    (3) Aggregate Sizes & Percentages   Size   Pct.

                                                               No.1   %

                                                               No.2   %

                                                                No.3   %

                                                                No.4   %

    (4) Aggregate
       Description:
       Color:

Source:

Producer:

D. Finish plaster

Number:

by (Manufacturer):

E. Sealer

Name:

by (Manufacturer):

10.10.9 Murals

The architect or his representative shall transfer design pattern shown on large scale drawings to base which receives finish plaster mural. The transferred design shall serve as guide for the application of flexible or rigid metal separating screeds. Artist's rendering, and color and aggregate key, shall be provided contractor at time of bidding.

NOTES

This specification reference presupposes that basic lathing and plastering specifications have already been prepared for the project and relates only to materials and application methods for finish plaster.

NOTE 1

Specify in lathing section control joints, casing beads, parting screeds, or other metal sections to define panels as shown on drawings, and to serve as grounds for finish plaster. Finish plaster panels should not exceed 120 square feet, or 11 feet in any direction. Specify total ground thickness and show on drawings.

NOTE 2

E.g. Portland Cement; Lime; Sand; Gypsum, etc.

NOTE 3

Other natural aggregate such as quartz, cinders, sea shells, or integrally colored manufactured aggregate, such as crushed glass, china, ceramics, may be specified providing they are weather resistant, permanent in color, moderately hard (3 or more on the MOH scale) and are compatible with the bedding coat.

NOTE 4

Liquid bonding agent may be applied to plaster basecoat before application of bedding coat. If bonding agent is desired, so specify; or permit use at option of contractor.

NOTE 5

Sealer is recommended to improve and retain color and cleanliness of finish plaster.

NOTE 6

Specify those bases applicable to project.

NOTE 7

See 7.1.2 Masonry Surfaces; 7.1.3 Poured Concrete Surfaces, Calif. Reference Specifications.

NOTE 8

Check with local building officials for method of application approved. All building codes require minimum overall thickness of exterior Portland cement plaster to be 7/8 inch; gypsum plaster minimum 1/2 inch.

NOTE 9

See 10.3. Specify pertinent information under Samples, 10.10.8.

NOTE 10

Overall thickness will depend on thickness of basecoat plaster plus thickness of bedding coat.

PART III EXECUTION

3.01 EXAMINATION

A. Prior to starting lathing or plastering work, carefully inspect installed work of other trades to verify that work is complete to the point where work of this section may properly commence.

B. Notify the architect or proper authorities in writing of conditions detrimental to the proper and timely completion of the lathing and/or plastering work.

C. Do not begin installation until all unsatisfactory conditions are resolved.

D. A pre-construction meeting is recommended with the architect and/or owner, primary contractor and representatives responsible for the windows, framing, flashing, roofing, sealants, stucco and any other building components interfacing with the stucco.

NOTE: **The examination of conditions is the responsibility of all parties involved in the project.**

3.02 PERFORMANCE

A. The work shall be performed by a skilled and trained work crew.

NOTE: **Work shall be performed to create a rigid, secure, level or plumb finish surface, with line dimensions and/or contours as indicated in reference standards or project documents.**

B. Install specified products and/or systems in accordance with reference standards, manufacturer's recommendations and this Stucco Resource Guide, unless indicated otherwise in project documents.

C. Flashings shall be installed prior to start of lathing or may be required to be integrated at the time of lathing. Flashing is not the scope of work of the contractor responsible for the stucco assembly.

3.03 INSTALLATION OF STUCCO TRIM ACCESSORIES

NOTE: Refer to details and other portions of this Stucco Resource Guide for additional information.

A. Verify that substrate and work by other trades are complete to the point at which installation of trim accessories may properly commence.

NOTE: Do not begin work until unsatisfactory conditions are resolved.

B. Attachments shall be firm enough to hold trim accessories in place without misalignment during plastering.

NOTE: Flanges or attachment points of trim accessories shall be secured to substrate in accordance with requirements of manufacturers of approved fasteners. Space per manufacturer's directions.

It is recommended that PVC or aluminum reveals, when used in conjunction with 2-inch x 2-inch welded or woven wire lath, incorporate a strip of expanded metal lath over the solid reveal flanges.

C. It is recommended that trim accessories have a small top flange "lip" that goes over the stucco and/or flange that embeds into the stucco.

D. Zinc alloy or PVC is recommended if trim accessories are exposed to a high-salt environment.

E. Install individual trim-accessory sections to each other at end joints for accurate alignment.

F. Install trim accessories in a manner that ensures a true, level and plumb stucco surface, and moisture resistant.

NOTE: Accurate alignment of trim accessories may require shimming, limited to the thickness of stucco.

Sections of flashing or trim accessories that butt each other (at corners or another condition) need to be lapped, caulked, or have a strip of self-adhering membrane over the joints. This is to prevent moisture from getting to the building structure.

G. Install the trim accessories in accordance with the required thickness of stucco basecoat and finish coat requirements. (Refer to Plaster Thickness section)

NOTE: Depth (grounds) of trim accessory to be in accordance with the thickness of stucco. (Refer to Plaster Thickness Tables.) For example, 3/4-inch (19-mm) stucco requires a minimum ground of a casing bead to be 3/4 inch (19 mm).

H. Install the longest possible lengths of trim accessories. A minimum continuous section (length) of 7 feet (2 m) is recommended.

3.04 TRIM ACCESSORY JOINTS

NOTE: Installation to be in accordance with section 3.03-Installation of Stucco Trim Accessories.

Trim accessory joints refer to various types: control joints, expansion joints, reveals, and/or any other devices or systems that divide (break) the stucco membrane surface.

Architect is to select the type of joint and indicate on drawing the location of joints.

Trim accessory joints provide relief of stresses from the structure and cement-plaster curing process.

Trim accessory joints provide for a plaster stop and a screed.

Aluminum and/or PVC reveals require that when the lath is installed over the flange, it totally covers it. The welded wire or woven wire lath shall be installed so as the crotch of the lath is over the flange.

It is recommended that PVC or aluminum reveals, when used in conjunction with 2 inch x 2-inch welded or woven wire lath, incorporate a strip of expanded metal lath over the reveal flanges.

A. The weather-resistant barrier must continue unbroken behind trim accessory joints in vertical or horizontal direction.

B. Locate to trim accessory joints strategically at points where building movement is anticipated.
   1. Wall penetrations
   2. Structural plate lines
   3. Junctures of dissimilar substrates
   4. Existing construction joints in structure
   5. Columns
   6. Cantilevered areas

C. Joints are recommended in stucco assemblies with lath reinforcement but have limited use in direct-applied stucco over concrete or concrete masonry surface.

D. It is recommended that trim accessory joints be weather-sealed by embedment in caulking at intersections when placed end-to-end and at the terminations.

E. It is recommended to install vertical joints continuously and to abut horizontal joints to vertical (be sure that weather-resistant barrier runs continuously behind joints).

NOTE: The use of horizontal reveals, flashing designs and/or other horizontal surface breaks may prevent continuous vertical joints.

Expansion joints govern over control joints.

F. Install longest possible lengths. No termination of a section within 24 inches (600 mm) of an intersection, with the exception of pre-manufactured trim accessory joint intersections.

G. It is recommended that trim accessory joints be installed on framed, sheathed construction so as to create stucco panels of smooth textures not more than 10 to 12 ft. in any direction (100 to 144 sq. ft. max.) or of coarse textures not more than 16 ft. in any direction (256 sq. ft. max.) in as square a configuration as possible.

NOTE: Maximum recommended length of a panel is 16 feet. Panel size approximately 2.5-to-1 ratio.

H. It is recommended that trim accessory joints be installed with concrete or concrete masonry construction so as to create a stucco assembly (with lath reinforcement) of 200 to 250 square feet (18 m² to 23 m²).

I. Installing control joints over continuous lath is an approved method.

NOTE: Control joint is a one-piece trim accessory.

Control joint may be installed in the field of panel only if there is framing behind it for attachment.

Control joints to be wire tied to lath with 18-gauge tie wire (as the preferred in all cases, and is recommended especially when lath goes over sheathing board other than wood).

J. The recommendation for the installation of expansion joints or reveals is to break the lath and lap it over on top of each of the flanges.

NOTE: Expansion joints may also serve as control joints (but not vice versa), if so specified and detailed.

Expansion joints and/or reveals may consist of one or two pieces.

Control joints are limited in their degree of movement. Two-piece expansion joints provide greater movement.

K. Sheathed framed construction with vertical trim accessory joints that require the lath to be terminated (cut) and installed on top of the flanges shall be placed at framing member locations. Lath shall be attached with appropriate fasteners through the trim accessory flange, sheathing and into the framing member.

NOTE: The lath/flange on each side of the trim accessory joint to be attached to a framing member. Double framing supports would be required at these locations.

This condition does not pertain to control joints installed over continuous lath.

3.05 LATHING FOR SOFFITS

A. Suspended soffits/ceilings shall be erected so that the finished basecoat cement plaster surface is true to line and level, with allowable tolerance of 1/4 inch (6 mm) in 10 feet (3.1 m)

NOTE: Construction documents shall indicate the approximate amount of deflection of structure to which suspended soffit is attached.

No attachment of soffit/ceiling abutting to walls or fascia (in order to prevent transfer of stresses).

Vertical compression struts to withstand upward wind pressures are not required unless indicated in project documents.

NWCB recommends a soffit framing system with sheathing (refer to 3.06-B) because it incorporates a gypsum sheathing board attached to furring "hat" channels. Most other authorities disagree with this use of sheathing under the lath on ceilings. They contend that it is unnecessary for flatness, and instead presents the possibility of trapping intruding moisture in the confined space, leading to condensation and accelerated oxidation of metal components.

Suspended ceiling systems shall comply with the requirements of CBC Table 25A-A and ASTM C 1063, Tables 2 and 3.

Refer to sections 3.03 and 3.04.

B. Spacing, attachment, size and type of hangers, fasteners and inserts shall be as required in reference standards in section 1.04.

NOTE: Fasteners and inserts shall have up-to-date testing information from manufacturers.

3.06 LATHING SYSTEMS FOR SUSPENDED SOFFITS (FOR DETAILS, SEE SOFFIT SECTION)

NOTE: For additional suspended soffit configurations (size, spans and spacing), refer to reference standards in section 1.04. (ASTM C 1063)

Construction documents to indicate either the soffit framing system (refer to 3.06 A).

A. Soffit framing system
  1. Hanger wires maximum support 12 square feet (1.5 m^2)
  2. 1½-inch (38-mm) main runner channels O.C. spacing 48 inches (1.2 m), hangers 36 inches O.C.

3. ¾-inch (19-mm) cross-furring channels O.C. spacing 16 inches (325 to 400 mm); if lesser weight lath is used, closer spacing is required.

4. 3.4-pound-per-square-yard diamond-mesh lath

5. Wire-tie lath to cross-furring channels with 18-gauge tie-wire, 6 inches (150 mm) O.C.

3.07 SOFFIT LATHING SYSTEM TO WOOD JOIST

A. 3/8-inch (10 mm) rib lath attached 6 inches (150 mm) O.C., with one of the following:

1. Roofing nails: 11 gauge. 7/16-inch diameter head. l-inch long

2. Staples: 16 gauge. 3/4-inch crown. 1 1/4-inch long

3. Screws: type W . wafer head. 1-inch long

NOTE: Fasteners must penetrate into the wood a minimum of 3/4 inch (19 mm) both legs of staple must penetrate wood. Fasteners to be centered, minimum 3/8 inch (10 mm) from edge of wood joist.

Maximum spacing of wood joist 24 inches (600 mm) O.C.

3.08 APPLICATION OF STUCCO BASECOAT FOR SOFFITS

NOTE: Do not use the double-back method of applying the basecoats.

Verify trim accessories and lath are installed so as to provide proper screeds, thickness and alignment for plastering.

Do not begin plastering work until unsatisfactory conditions are resolved.

Cement plaster can be applied by trowel or machine.

Apply basecoat to entire horizontal surface and/or section without interruption or "cold joints."

A. Indicate the permissible mix number and its proportions to be used for the basecoat

NOTE: Delete basecoat mixes not applicable to this project.

Do not change selected mix or proportions after plastering operation has started.

Measurement of ingredients (materials of the mix) to be done with devices of known volume, accurate, and successive batches proportionally alike.

Use only enough mix water to make plaster a workable consistency.

Admixtures to be submitted for approval.

Basecoat admixtures designed for improving the mix or for enhancing the plaster application must have been manufactured strictly for that purpose.

Admixtures must be accurately measured and used only in accordance with manufacturer's recommendations.

B. Apply stucco first ("scratch") coat in a nominal thickness of 3/8" (10 mm) minimum, but 1/2" (11 mm) preferred. First coat to completely embed the lath. First coat to be thick enough to allow for scoring of cement plaster surface. (Refer to Plaster Thickness Tables)

NOTE: Scoring of the first coat should be uniform and shallow, approximately 1/8 inch (3 mm).

C. Moist cure the first ("scratch") coat until it is sufficiently rigid to accept the succeeding coat; otherwise, moist cure for a minimum of 48 hours then allow to dry completely, and dampen before application of the second ("brown") coat.

NOTE: First coat to be sufficiently rigid before receiving the second coat.

D. Apply stucco second ("brown") coat in a nominal thickness of 3/8 inch (10 mm) over stucco first coat. Second coat thickness to bring the combined basecoats (first and second) thickness to a nominal thickness of 3/4" to 7/8" inch. (Refer to Plaster Thickness Tables)

NOTE: Apply second coat over a damp first coat. If required, apply a fine spray of clean water, so as to dampen only. Do not saturate. Allow water sheen to disappear before applying the second coat.

E. Apply the second coat with sufficient material and pressure to ensure a tight uniform bond to the first coat.

NOTE: Apply second coat so as not to deform or crack the first coat.

F. Rod the second coat to a true, even plane, filling surface defects with cement plaster.

G. Float the second coat surface uniformly.

NOTE: The floating process densifies the basecoat and provides a proper surface for the finish coat application.

Float the basecoat after it has set and when moisture is still present in it.

Floating the basecoat that is to receive an acrylic finish coat is critical because of the thickness of this finish.

3.09 CURING OF SOFFIT BASECOAT – REFER TO SECTION 3.40

NOTE: After the second coat has been applied, moist cure it as required, and then wait a minimum of seven days before the start of finish coat application.

3.10 STUCCO FINISH COAT – REFER TO SECTION 3.41

3.11 ACRYLIC FINISH COAT – REFER TO SECTION 3.42

3.12 ADDITIONAL TYPES OF FINISHES – REFER TO SECTION 3.43

3.13 CEMENT PLASTER DIRECT TO CONCRETE

NOTE: Concrete to be cured for a minimum of 28 days. 37 Northwest

Not recommended for horizontal surfaces (other than ceilings/soffits).

Concrete surface to be straight and true, in accordance with standards.

Concrete surface to be clean, no form release agents, no curing compounds or other elements in concrete surface preventing a proper bond.

Concrete should be in good condition and have uniform absorption rate over entire surface.

Concrete in poor condition (spalling, delamination, voids) or other surface defects requires repair before plastering.

If the concrete surface is not in proper condition, self-furring lath must be installed.

If there is any concern about the bondability of the cement plaster direct to concrete, a test area is recommended. Apply ¼" - ½" thick, by 1' x 1' of cement plaster basecoat to concrete; let it cure for 3 - 4 days. Then try to remove it. Do this type of test at more than one location.

Use a liquid bonding agent approved for exterior applications, following manufacturer's instructions and recommendations, if needed.

Acrylic finish coat is not recommended for concrete retaining walls.

3.14 STUCCO "FINISH COAT" DIRECT TO CONCRETE

NOTE: The finish coat over concrete is used to enhance the surface appearance; it is not designed to "true up" the surface.

Smooth troweled stucco finishes are not recommended because of the possibility of very fine map cracking. If smooth troweled finish is specified, surface should be painted with a high-quality acrylic exterior paint.

> A. Apply liquid bonding agent directly to concrete surface in accordance with the recommendation of manufacturer of the material.
>
> B. Application of the stucco finish coat-Refer to section 3.41.

3.15 ACRYLIC FINISH COAT DIRECT TO CONCRETE – Refer to section 3.42.

NOTE: Do not apply finish until all irregularities on the concrete surface have been corrected.

3.16 TWO COAT STUCCO DIRECT TO CONCRETE – Refer to "Stucco Wall Assemblies" section detail A7 in this Stucco Resource Guide.

NOTE: This method provides for an economical way of creating an acceptable level finished wall surface.

The system provides a leveling basecoat (brown coat) over the concrete surface and makes a suitable base for the finish coat with limited total stucco thickness of 1/2 inch (13 mm).

Refer to section 3.13..38 Northwest

> A. Apply liquid bonding agent directly to concrete surface only in accordance with the recommendation of manufacturer of the material.
>
> B. Install trim accessories-termination trim accessory, corner reinforcements (corner beads), accessory joints (control joints) and other specified accessories in accordance with sections 3.03 and 3.04.
>
> **NOTE:** Install per related detail sections of this Stucco Resource Guide.
>
> Delete trim accessories not applicable.
>
> Depth (grounds) of trim accessory to be in accordance with the thickness of stucco. (Refer to Plaster Thickness Tables)
>
> Install so as to provide the required screed and ground and to create a true and level surface. (Refer to Plaster Thickness Tables)
>
> C. Trim accessory attachment

NOTE: The installation of fasteners and their pullout resistance shall be in accordance with the manufacturer's recommendations and safety requirements.

If there is any question on the pullout strength of the fasteners, sample testing on the wall is recommended.

The psi of the concrete, how long the concrete has cured and the physical condition of the concrete affect the selection of proper fasteners for the attachment of trim accessories and lath.

Fastener head shall be capable of lapping over two or more strands (recommended to be placed at the crotch of the trim accessory flange) and/or lapped over attachment hole.

D. Indicate type of stucco termination trim accessory (casing bead): Indicate name of manufacturer (optional):

E. Attach termination trim accessory to concrete surface with hardened concrete stub nails; low-velocity, power-actuated pins or drill-and-drive fasteners. Fastener heads minimum of 3/8-inch (10-mm) diameter, length of fasteners 3/4 inch (19 mm); spacing of fasteners 8 to 12 inches (200 to 300 mm) O.C.

F. Indicate the type of corner reinforcement:

Indicate name of manufacturer (optional):

NOTE: Corner reinforcement is recommended at vertical exterior and slanted corners.

G. Attach corner reinforcement to concrete surface with hardened concrete stub nails; low-velocity, power-actuated pins or drill-and-drive fasteners. Fastener heads minimum of 3/8-inch (10-mm) diameter, length of fasteners 3/4 inch (19 mm); spacing of fasteners 12 inches (300 mm) O.C. staggered.

H. Indicate the type of trim accessory joint (control joint, expansion joint, reveal or other style):

Indicate name of manufacturer (optional):

NOTE: The use of trim accessory joints (control joints) is limited on a stucco system direct to concrete masonry, and they are not required as frequently as for framed construction.

Control or expansion joints are recommended at locations of concrete expansion joints (construction joints).

Control joints are recommended if the area exceeds 250 square feet (23 m^2). Panel should be in as square a configuration as possible.

The location of trim accessory joints and the type are by the architect.

I. Attach trim accessory joints to concrete surface with hardened concrete stub nails; low-velocity, power-actuated pins or drill-and-drive fasteners. Fastener heads minimum of 3/8-inch (10-mm) diameter, length of fasteners 3/4 inch (19 mm); spacing Of fasteners 12 inches (300 mm) O.C. staggered.

J. Application of stucco basecoat

NOTE: Inspect concrete surface to ensure it is in proper condition for a direct application of cement plaster.

Do not begin plastering work until unsatisfactory conditions are resolved.

Cement plaster can be applied by trowel or machine.

Apply basecoat to entire wall panel and/or section without interruption or "cold joints."

K. Indicate the permissible mix and its proportions to be used for the basecoat

   NOTE: Delete basecoat mixes not applicable to this project.

   Do not change selected mix or proportions after plastering operation has started.

   Measurement of ingredients (materials of the mix) to be done with a device of known volume. Measure accurately. Make successive batches proportionally alike.

   Use only enough mix water to make plaster a workable consistency.

   Admixtures are for improving the basecoat mix or enhancing the plaster application. Use products manufactured strictly for this purpose and in accordance with manufacturer's recommendations.

L. Apply basecoat (brown coat) in a nominal thickness of 3/8 inch (10 mm). Refer to Plaster Thickness Tables.

   NOTE: Basecoat direct to concrete 3/8 inch to 1/2 inch (10 to 13 mm) thick, not greater than 1/2 inch (13 mm)

M. Apply basecoat with sufficient material and pressure to ensure tight contact with concrete surface and uniform thickness.

N. Rod the basecoat to a true and even plane, filling basecoat surface defects with cement plaster.

O. Float the basecoat surface uniformly.

   NOTE: The floating process densifies the basecoat and provides a proper surface for the finish coat application.

   Float the basecoat after it has set and when moisture is still present in it. (The float should not adhere to the surface that is to be worked)

3.17 CURING OF BASECOAT – Refer to section 3.40.

3.18 STUCCO FINISH COAT – Refer to section 3.41.

3.19 ACRYLIC FINISH COAT – Refer to section 3.42.

3.20 ADDITIONAL TYPES OF FINISHES – Refer to section 3.43.

3.21 TWO-COAT STUCCO DIRECT TO CONCRETE MASONRY "CMU"

NOTE: This method consists of a leveling basecoat (brown coat) followed by a finish coat.

Most concrete masonry is an excellent substrate for direct-applied cement plaster. If the units lack porosity, a liquid bonding agent approved for exterior use should be used, to ensure bond.

The two-coat stucco system trues up the surface, creating a weather-resistant membrane and pleasing finished surface.

This system is recommended and commonly used over a concrete masonry substrate or other similar types of surfaces that will develop the requirement.

Reinforcement lath is omitted in this system. Do not tool the mortar; joints of the concrete masonry assembly. It is recommended to have mortar joints struck flush with the surface.

Concrete masonry wall must have been cured in accordance with industry standards (a minimum of 30 days) before application of cement plaster basecoat and carry the designed dead loads before plastering.

Concrete masonry surface to be clean, no substances on the surface or in the units and/or joints which would prevent a proper bond with the stucco basecoat.

Coated (painted) concrete masonry surfaces require self-furring lath attached in accordance with standards or the removal of the coating.

The degree of the masonry wall misalignment will dictate either the direct application of a two-coat stucco system or the use of self-furring lath in order to achieve a true wall surface plane.

Concrete masonry assembly shall be constructed in accordance with its application standards.

   A. Install trim accessories-termination trim accessories (casing beads), corner reinforcements (corner beads); trim accessory joints (control joints) and other specified accessories according to sections 3.03 and 3.04 to concrete masonry substrate.

   NOTE: Install per related detail sections of this Stucco Resource Guide.

   Delete trim accessories not applicable to project.

   Depth (grounds) of trim accessory shall be in accordance with the thickness of stucco. (Refer to Plaster Thickness Tables)

   Install so as to provide the required screed and ground in accordance with a nominal stucco thickness of 1/2 inch (13 mm) and to create a true and level surface. (Refer to Plaster Thickness Tables)

   B. Trim accessory attachment

   NOTE: The installation of fasteners and their pullout-resistance requirements shall be in accordance with the manufacturer's requirements and safety recommendations.

   If there is any question on the pullout strength of the fasteners, sample testing on the wall is recommended.

   Fastener head shall be capable of lapping over two or more strands (recommended to be placed at the crotch of the trim accessory) and/or lapped over attachment hole.

   Spacing or fasteners depends on type and size of the concrete masonry units.

   For substrates of concrete masonry units, brick or tile, it is recommended that the fasteners be placed in the mortar joints. This will minimize damage to the individual units.

   C. Indicate types of termination trim accessories (casing beads):

   Indicate name of manufacturer (optional):

   D. Attach termination trim accessory to concrete masonry surface with hardened concrete stub nails; low-velocity, power-actuated pins or drill-and-drive fasteners.

   Fastener heads minimum of 3/8-inch (10-mm) diameter, length of fasteners 3/4 inch (19 mm); spacing of fasteners 8 to 12 inches (200 to 300 mm) O.C.

   E. Indicate the type of corner reinforcement:

   Indicate name of manufacturer (optional):

NOTE: Corner reinforcement is recommended at vertical exterior and slanted corners.

F. Attach corner reinforcement to concrete masonry surface with hardened concrete stub nails; low-velocity, power-actuated pins or drill-and-drive fasteners. Fastener heads minimum of 3/8-inch (10-mm) diameter, length of fasteners 3/4 inch (19mm); spacing of fasteners 12 inches (300 mm) O.C. staggered.

G. Indicate the type of trim accessory joint (control joint, expansion joint, reveal or other style):

Indicate name of manufacturer (optional):

NOTE: Control or expansion joints are recommended at locations of concrete masonry expansion joints (construction joints).

The use of trim accessory joints (control joints) is limited on a stucco system direct to concrete masonry and are not recommended as frequently as for framed construction.

Control joints are recommended if the area exceeds 250 square feet (23 m^2). Panel should be in as square a configuration as possible.

The location of trim accessory joints and the type are by the architect.

H. Attach trim accessory joint to concrete masonry surface with hardened concrete stub nails; low-velocity, power-actuated pins or drill-and-drive fasteners. Fastener heads minimum of 3/8-inch (10-mm) diameter, length of fasteners 3/4 inch (19 mm); spacing of fasteners 12 inches (300 mm) O.C. staggered.

I. Application of stucco basecoat

NOTE: Inspect concrete masonry surface to ensure it is in proper condition for a direct application of cement plaster.

Do not begin plastering work until unsatisfactory conditions are resolved.

Cement plaster can be applied by trowel or machine.

Apply basecoat to entire wall panel and/or section without interruption or "cold joints."

J. Indicate the permissible mix number and its proportions to be used for the basecoat:

NOTE: Delete basecoat mixes not applicable to this project.

Do not change selected mix or proportions after plastering operation has started.

Measurement of ingredients (materials of the mix) to be done with a device of known volume. Measure accurately. Make successive batches proportionally alike.

Use only enough mix water to make plaster a workable consistency.

Admixtures are for improving the basecoat mix or enhancing the plaster application. Use products manufactured strictly for this purpose and in accordance with manufacturer's recommendations.

K. Apply basecoat (brown coat) in a nominal thickness of 1/2 inch (13 mm). Refer to Plaster Thickness Tables.

NOTE: Basecoat over liquid bonding agent or direct to concrete masonry surface 3/8 inch (10 mm) to 5/8 inch (13 mm) thick, not greater than 5/8 inch (13 mm).

L. Apply basecoat with sufficient material and pressure to ensure tight contact with concrete masonry surface and uniform thickness.

NOTE: Dampen the substrate by spraying with clean water just before starting plastering. Dampening of the surface is not required if a liquid bonding agent has been used.

It is advisable first to apply a dash bond coat or liquid bonder to the concrete masonry surface in order to uniform the suction and help prevent the telegraphing of the mortar joints. Dash coat does not replace one of the specified number of coats.

M. Rod the basecoat to a true, even plane, filling basecoat surface defects with cement plaster.

N. Trowel-float the basecoat surface uniformly.

NOTE: The floating process densifies the basecoat and provides a proper surface for the finish coat application.

Float the basecoat after it has set and when moisture is still present in it. (The float should not adhere to the surface that Is to be worked)

3.22 CURING OF BASECOAT – Refer to section 3.40.

3.23 STUCCO FINISH COAT – Refer to section 3.41.

3.24 ACRYLIC FINISH COAT – Refer to section 3.42.

3.25 ADDITIONAL TYPES OF FINISHES – Refer to section 3.43.

3.26 STUCCO ASSEMBLY ATTACHED TO CONCRETE

NOTE: This system is recommended if a greater thickness of stucco is necessary or if the concrete surface will not provide an adequate bond for direct application of cement plaster.

Concrete surface to be in good condition, no large voids, no spalling and no delamination. Concrete to be true and straight.

Surface to be within tolerance standards for a concrete tilt-up or poured-in place walls.

Concrete to be cured a minimum of 30 days before start of any stucco work.

Do not begin installing trim accessories or lath until all unsatisfactory conditions are resolved.

A weather-resistant barrier is not required for this stucco assembly.

A. Install trim accessories-termination trim accessories (casing beads), corner reinforcements, trim accessory joints (control joints) and other specified accessories in accordance with sections 3.03 and 3.04.

NOTE: Install per related detail sections of this Stucco Resource Guide.

Delete trim accessories not applicable to project.

Depth (grounds) of trim accessory shall be in accordance with the thickness of stucco. Refer to Plaster Thickness Tables.

Install so as to provide the required screed and ground in accordance with a nominal stucco thickness of 3/4 inch (19 mm) and to create a true and level surface. (Refer to Plaster Thickness section)

Install per manufacturer's recommendations.

B. Trim accessory and lath attachment

   NOTE: The installation and pullout resistance of fasteners shall be in accordance with manufacturer's recommendations and safety requirements.

   Fasteners shall be placed in the crotch of the flange and/or the crotch of the lath.

   The psi of the concrete, how long the concrete has cured and the physical condition of the concrete affect the selection of proper fasteners for the attachment of trim accessories and lath.

   If there is any question on the pullout strength of the fasteners, sample testing on the wall is recommended.

   Bend lath and continue around corners, a minimum of 12 inches (300 mm).

   Apply so as long dimension of lath is horizontal. Apply lath taut.

   Attachment of lath should be at furring points.

C. Indicate type of stucco termination trim accessory (casing bead):

   Indicate name of manufacturer (optional):

D. Attach termination trim accessory to concrete surface with low-velocity, power-actuated pins or drill-and-drive fasteners. Fastener heads minimum of 3/8-inch (10-mm) diameter, length of fasteners 3/4 inch (19 mm); spacing of fasteners 8 to 12 inches (200 to 300 mm) O.C.

E. Indicate the type of corner reinforcement:

   Indicate name of manufacturer (optional):

   NOTE: Selection of proper type in accordance with specified finish coat.

   A welded-wire plastic-nose or PVC corner reinforcement is recommended if finish coat is a sand-finish stucco or acrylic finish coat.

   Corner reinforcement is recommended at vertical exterior and slanted corners.

F. Attach corner reinforcement to concrete surface with low-velocity, power-actuated pins or drill-and-drive fasteners. Fastener heads minimum of 3/8-inch (10-mm) diameter, length of fasteners 3/4 inch (19 mm); spacing of fasteners 12 inches (300 mm) O.C. staggered, or attach to lath with 18-gauge tie wire, spaced 12 inches (300 mm) O.C. staggered.

G. Indicate the type of trim accessory joint (control joint, expansion joint, reveal or other style):

   Indicate name of manufacturer (optional):

   Control joints are recommended for areas greater than 200 square feet (18 m^2). Maximum recommended length of a panel is 20 feet (6 m). Panel size should not exceed a 3-to-1 ratio.

   The location of trim accessory joints and the type are by the architect.

H. Attach trim accessory joints to concrete surface with low-velocity, power-actuated pins or drill-and-drive fasteners. Fastener heads minimum of 3/8-inch (10-mm) diameter, length of fasteners 3/4 inch (19 mm); spacing of fasteners 12 inches (300 mm) O.C.

NOTE: Attachment method depends on type and function of trim accessory joints.

The recommendation is for the control joints be installed over the continuous lath, except at specified locations requiring an expansion-type trim accessory joint.

I. Installation of expansion joints requires lath to be cut and attached to both sides of the expansion joint flange staggered 12 inches (300 mm) O.C., attach with wire ties or appropriate fasteners into concrete surface. (Control joints may be installed on top of the lath.)

NOTE: Expansion joints are recommended where there are concrete expansion joints (construction joints).

J. Install galvanized self-furring diamond-mesh metal lath, 2.5 or 3.4 pounds per square yard (1.4 kg/m 2 or 1.8 kg/m^2) to vertical concrete surface (walls). For horizontal concrete surfaces (ceilings), use 3.4-pound-per-square-yard (1.8 kg/m^2) self-furring lath.

NOTE: Lath shall be applied with the long dimension of sheets horizontal.

Diamond-mesh lath is recommended rather than other styles of stucco reinforcement lath because of the smaller size of opening in the lath provides for easier and more positive securing of fasteners into a concrete surface.

K. Attach lath to concrete surface, using low-velocity, power-actuated pins or drill-and-drive fasteners. Fastener heads minimum of 3/8-inch (10-mm) diameter, length of fasteners 3/4 inch (19 mm).

L. Spacing of fasteners for 2.5-pound-per-square-yard lath to walls shall be 16 inches (400 mm) O.C. horizontally and 7 inches (180 mm) O.C. vertically. Spacing of fasteners for 3.4-pound-per-square-inch lath to walls shall be 24 inches (600 mm) O.C. horizontally and 6 inches (180 mm) vertically. Spacing of fasteners for lath to ceilings shall be 16 inches (400 mm) and 7 inches (180 mm) O.C.

3.27 STUCCO ASSEMBLY ATTACHED TO CONCRETE/APPLICATION OF STUCCO BASECOAT – Refer to section 3.39 in this Stucco Resource Guide.

3.28 CURING OF BASECOAT – Refer to section 3.40.

3.29 STUCCO FINISH COAT – Refer to section 3.41.

3.30 ACRYLIC FINISH COAT – Refer to section 3.42.

3.31 ADDITIONAL TYPES OF FINISHES – Refer to section 3.43.

3.32 STUCCO ASSEMBLY ATTACHED TO CONCRETE MASONRY – Refer to "Stucco Wall Assemblies"

NOTE: This system is used if a greater thickness of stucco is required than is recommended for stucco directly applied or if the concrete masonry surface will not provide an adequate bond with the cement plaster.

A concrete masonry wall must have been cured in accordance with industry standards (a minimum of 30 days) before application of stucco assembly.

A concrete masonry surface is to be in good condition, no spalling or delamination, and no large voids.

Do not tool the mortar joints of the concrete masonry assembly. It is recommended to have mortar joints struck flush with the surface.

A concrete masonry substrate shall be aligned in accordance with its application standards.

    A. Install trim accessories-termination trim accessories (casing beads), corner reinforcements (corner beads); trim accessory joints (control joints) and other specified accessories in accordance with sections 3.03 and 3.04.

    NOTE: Install per related detail sections of this Stucco Resource Guide.

    Delete trim accessories not applicable to project.

    Install per manufacturer's recommendations.

    Depth (grounds) of trim accessory to be in accordance with the thickness of stucco. (Refer to Plaster Thickness Tables)

    Install so as to provide the required screed and ground in accordance with a nominal stucco thickness or 3/4 inch (19 mm) and to create a true and level-surface. (Refer to Plaster Thickness Tables)

    B. Trim accessory and lath attachment

    NOTE: The installation and pullout resistance of fasteners shall be in accordance with manufacturer's recommendations and safety requirements.

    Fasteners shall be placed in the crotch of the trim accessory flange and/or the crotch of the lath.

    The type and condition of the concrete masonry surface affect the selection of proper fasteners for the attachment of trim accessories and lath.

    If there is any question on the pullout strength of the fasteners, sample testing on the wall is recommended.

    Bend lath and continue around corners, a minimum of 12 inches (300 mm), or install a corner reinforcement.

    Apply so as long dimension or lath is horizontal. Apply lath taut.

    Attachment of lath should be at furring points. ,

    Fasteners shall be placed in the crotch of the flanges and or the crotch of the lath.

    Spacing of fasteners may depend on type and size of the concrete masonry units.

    For substrates of concrete masonry units, brick or tile, it is recommended that the fasteners be placed in the mortar joints. (This will minimize damage to the individual units).

    C. Indicate type of termination trim accessory (casing bead/foundation weep screed):

    Indicate name of manufacturer (optional):

    D. Attach termination trim accessory to concrete masonry surface with hardened concrete stub nails; low-velocity, power-actuated pins or drill-and-drive fasteners.

    Fastener heads minimum of 3/8-inch (10-mm) diameter, length of fasteners 3/4 inch (19 mm); spacing of fasteners 12 inches (300 mm) O.C.

E. Indicate the type of corner reinforcement:

and name of manufacturer (optional):

NOTE: Corner reinforcement is recommended at vertical exterior and slanted corners.

Selection of proper type in accordance with specified finish coat.

A welded wire plastic-nose or PVC corner reinforcement is recommended if finish coat is a sand-finish stucco or acrylic finish coat.

F. Attach corner reinforcement to concrete masonry surface with hardened concrete stub nails; low-velocity, power-actuated pins or drill-and-drive fasteners. Fastener heads minimum of 3/8-inch (10-mm) diameter, length of fasteners 3/4 inch (19 mm); spacing of fasteners 12 inches (300 mm) O.C. staggered.

G. Indicate type of trim accessory joint (control joint, expansion joint, reveal or other style):

Indicate name of manufacturer (optional):

NOTE: Control or expansion joints are recommended at locations of concrete masonry expansion joints (construction joints).

Trim accessory joints (control joints) are recommended, but not as frequently as for framed construction.

Control joints are recommended for areas greater than 200 square feet.

Maximum recommended length of a panel is 20 feet (6 m). Panel size should not exceed a 3-to-1 ratio.

The location of trim accessory joints and the type are by the architect.

H. Attach trim accessory joints to concrete masonry surface with hardened concrete stub nails; low-velocity, power-actuated pins or drill-and-drive fasteners. Fastener heads minimum of 3/8-inch (10-mm) diameter, length of fasteners 3/4 inch (19 mm); spacing of fasteners 12 inches (300 mm) O.C. staggered.

I. Installation of expansion joints requires lath to be cut and attached to both sides of the expansion joint 12 inches (300 mm) O.C. staggered, with appropriate fasteners, into concrete masonry joints.

J. Installation of lath reinforcement to concrete masonry

NOTE: Do not begin lathing work until all unsatisfactory conditions are resolved.

A weather-resistant barrier is not normally required or recommended for this stucco assembly. A combination self-furring lath and weather-resistant barrier (paper-backed lath) is the style of lath recommended if the project documents specify a weather-resistant barrier over the concrete masonry substrate.

Lath shall be applied with the long dimension of rolls and/or sheets horizontal.

K. Indicate type of self-furring lath:

Indicate name of manufacturer (optional):

Refer to section 2.07 for lath selection.

NOTE: Delete lath types not applicable.

If type of lath is not specified, it is the contractor installing the lath who selects the type for the project. Lath selected by contractor is to be installed only after it has been approved.

All lath to be in accordance with this Stucco Resource Guide and its reference standards. Refer to section 1.04.

L. Attach lath to concrete masonry surface with hardened concrete stub nails; low-velocity, power-actuated pins or drill-and-drive fasteners. Fastener heads minimum of 3/8-inch (10-mm) diameter, length of fasteners 3/4 inch (19 mm).

M. Spacing of fasteners for attachment of woven wire lath, welded wire lath and 2.5-pound-per-square-yard diamond-mesh lath: 16 inches (400 mm) O.C. horizontally.

For attachment of 3.4-pound-per-square-yard diamond-mesh lath: 24 inches (600 mm) O.C. horizontally. Vertical attachment of lath shall be 7 inches (180 mm), 6 inches UBC.

3.33 STUCCO ASSEMBLY ATTACHED TO CONCRETE MASONRY/APPLICATION OF STUCCO BASECOAT – Refer to section 3.39 in this Stucco Resource Guide.

3.34 CURING OF BASECOAT – Refer to section 3.40.

3.35 STUCCO FINISH COAT – Refer to section 3.41.

3.36 ACRYLIC FINISH COAT – Refer to section 3.42.

3.37 ADDITIONAL TYPES OF FINISHES – Refer to section 3.43.

3.38 STUCCO ASSEMBLY TO SHEATHED CONSTRUCTION – Refer to "Stucco Wall Assemblies" section in this Stucco Resource Guide.

NOTE: Framed construction refers to a wood framing assembly or a light-gauge steel framing assembly with appropriate sheathing.

It is recommended that the minimum size of wood framing member the stucco system is going over be 2 x 4 and standard or better grade.

It is recommended that the minimum size of steel framing members be 3-1/2 inches (88 mm) and a minimum of 20-gauge (.0329)

Open-frame construction is an approved framing method for attachment of stucco assembly, but far better results are achieved when rigid sheathing is employed.

Exterior sheathing board to have firm contact with framing members (no gaps), and wood-based sheathing panels shall be spaced 1/8" on ends and edges to allow for expansion.

Framing assembly (substrate) to be true, straight and level. Align in accordance with its application standards.

It is recommended that the wood framing and wood sheathing board have a moisture content of less than 19 percent before starting the scope of the stucco work.

Stucco is classified as a brittle exterior cladding, which means the cement plaster membrane is not flexible. The substrate to which the stucco systems is attached must be rigid and a minimum design deflection of L/360, although the code allows L/240.

The structure of the building to which the stucco system is installed can be load or non-load bearing walls.

It is recommended that the shear value of the wall to be created by the structure itself, not the stucco system.

Appropriate exterior sheathing boards over which to install the stucco assembly:

1. Exterior gypsum sheathing
2. Glass-mat gypsum sheathing
3. Cementitious backer board (may require paperback lath)
4. Exterior-grade plywood
5. Oriented strand board
6. Rigid expanded polystyrene (EPS)

Exterior-grade plywood and oriented strand board shall be installed with a minimum of 1/8 inch (3 mm) gap along all panel edges and ends.

Exterior sheathing boards to be installed per the manufacturer's recommendations and building code requirements.

Exterior sheathing board, once installed, requires protection from climatic conditions, until the installation of the stucco system.

Do not begin stucco system work until all unsatisfactory conditions are resolved.

Sections of flashing and trim accessories that butt each other (at comers or another condition) need to be lapped, caulked, and have a strip of self-adhering membrane over the joints. This is to prevent moisture from getting to the building structure.

A. The stucco contractor is to install stucco system component flashing (window head flashing, expansion joint/weep flashing) only if this scope of work is indicated in the project construction documents and stated in the stucco contractors performance contract.

   NOTE: Stucco system component flashing does not refer to standard roof or other types of building flashing.

   Refer to the section details and "Scope of Work Responsibilities" section of this Stucco Resource Guide.

   Delete if not applicable to project.

B. Install a weather-resistant barrier "WRB" as part of the stucco assembly. Weather-resistant barrier, per F.S. UU-B-790A grade D 60-minute kraft paper; spunbounded olefin housewrap for a stucco system, or an approved weather-resistant barrier designed for a stucco system.

   NOTE: It is required to use weather-resistant barriers with all wood or steel framed structures.

   This guide requires two layers of Grade D 60-minute building paper or two layers of olefin housewrap over wood-based sheathing. A combination of these two weather-resistant barriers can be used. The first layer is to be the housewrap weather-resistant barrier.

It is recommended that one layer of weather-resistant barrier be installed over all other types of sheathing boards.

All flashing and weather-resistant barrier to be installed in such a manner so as to prevent moisture from entering at all edges (tops and sides).

Do not damage, in any way, the weather-resistant barrier during installation. If damage occurs, repair before starting the trim and lathing work and/or completely replace the weather-resistant barrier.

The weather-resistant barrier is to be installed "shingle-fashion" so that natural direction of water flow would be over and onto the next sheet.

Install long dimension horizontal to framing.

Weather-resistant barrier shall have horizontal laps of 2 inches (50 mm) minimum. Vertical laps shall be 6 inches (150 mm) minimum.

C. Attach weather-resistant barrier to sheathing with small staples so as it is taut and flat.

Refer to details in the "Window Flashing Applications" section of this Stucco Resource Guide.

D. Install stucco system trim accessories-foundation weep screed, casing beads, corner reinforcements, trim accessory joints (control joints, expansion joints, reveals) and other specified accessories in accordance with sections 3.03 and 3.04.

NOTE: Install per related detail sections of this Stucco Resource Guide.

Delete trim accessories not applicable to project.

Install per manufacturer's recommendations.

The depth (grounds) of trim accessory to be in accordance with the thickness of stucco. (Refer to Plaster Thickness Tables)

Install so as to provide the required screed and ground in accordance with nominal stucco thickness requirements. (Refer to Plaster Thickness section.)

Install so as to produce a true and level surface for the cement plaster.

E. Trim accessory and lath attachment

NOTE: The type and condition of the sheathing board and framing members affect the selection of proper fasteners for the attachment of trim accessories and lath.

The installation and pullout resistance of fasteners shall be in accordance with manufacturer's recommendations and safety requirements.

Fasteners shall be placed in the crotch of the trim accessory flange and/or the crotch of the lath openings.

If there is any question on the pullout strength of the fasteners, sample testing on the wall is recommended.

The same fasteners can be used as a combination attachment for trim accessories and lath (and in some construction, the sheathing board).

Lath shall be applied with long dimension of the sheets or rolls horizontal (perpendicular) to framing members. Apply lath taut.

Bend lath and continue around corners to next framing member or install external corner reinforcement.

Attachment of lath should be at furring points.

Thickness of sheathing board or boards dictates the length of fasteners.

Fasteners to penetrate into wood or steel framing members. Attachment into only the sheathing board is not acceptable.

Fasteners to penetrate a minimum of 3/4 inch (19 mm) into wood framing members.

The thickness of sheathing boards of plywood and oriented strand board does not make up part of the minimum 3/4-inch (19-mm) fastener penetration requirement.

Fasteners to be centered on flange (ends) or framing member. A minimum of 3/8 inch (10 mm) from edge. Both legs of staple to penetrate framing member.

Fasteners to penetrate a minimum of 2 full threads past the steel framing member flange.

All types of fasteners shall bear tight against attachment item.

Construction methods and design (layout) may dictate that some fasteners only penetrate into sheathing board in order to attach trim accessory.

Sheathed framed construction with vertical trim accessory joints that require the lath to be terminated (cut) and installed on top of the flanges shall be placed at framing member locations. Lath shall be attached with appropriate fasteners through the trim accessory flange, sheathing and into the framing member. The lath/flange on each side of the trim accessory joint is recommended to be attached to a framing member. Double framing supports may be required at these locations. This condition does not pertain to control joints installed over continuous lath.

F. Indicate type of termination-at-foundation trim accessory (foundation weep screed, as required by Building Code):

Indicate name of manufacturer (optional): '

NOTE: Refer to "Termination at Foundation" detail section of this Stucco Resource Guide.

1. Foundation weep screeds and in some design details casing beads can serve as a termination for the stucco at the building foundation area with the local code jurisdiction's approval.

Building codes require a foundation weep screed.

Finished edge of stucco surface to be no lower than 4 inches (100 mm) from finished grade or 2 inches above a hard surface.

G. Fasteners for attachment of trim accessories, foundation trim and lath

NOTE: Refer to section 3.38-E.

Length of fasteners depends on thickness of sheathing boards.

Stucco contractor has the option of selecting appropriate fasteners from list below.

Wood-sheathed framing:

1. Roofing nails: 11-gauge . 7/16-diameter head. 1 1/2 inches long (i.e., 1/2 inch wood sheathing)

2. Staples: 16-gauge . 3/4-inch crown. 1 3/8 long i.e., 1/2 inch wood sheathing)

3. Type W screws: wafer head. 1 1/2 inches long (i.e., 1/2 inch wood sheathing)

Sheathed steel framing:

1. Type S screws: wafer head. self-drilling. 1 inch (25 mm) long (i.e., 1/2 inch sheating)

H. Indicate type of foundation trim accessory (foundation weep screed or casing bead):

Indicate name of manufacturer (optional):

I. Attach foundation trim accessory to building structure, with appropriate fastener selected from list in section 3.38-G. Spacing of fasteners 12 inches (300 mm) O.C.

NOTE: Refer to the "Termination at Foundation" detail section of this Stucco Resource Guide.

J. Indicate type of termination trim accessory (casing bead):

Indicate name of manufacturer (optional):

K. Attach termination trim accessory to building structure, with appropriate fastener selected from list in section 3.38-G. Spacing of fasteners 12 inches (300 mm) O.C.

NOTE: Refer to detail sections of this Stucco Resource Guide.

L. Indicate type of corner reinforcement (corner bead):

Indicate name of manufacturer (optional):

NOTE: Corner reinforcement is recommended at vertical exterior and slanted corners.

Selection of proper type in accordance with specified finish coat.

A welded wire plastic-nose or PVC corner reinforcement is recommended if finish coat is a sand-finish stucco or acrylic finish coat.

M. Attach corner reinforcement to building structure, with appropriate fastener selected from list in section 3.38-G, or attach to lath with 18-gauge tie wire. Spacing of fasteners 12 inches (300 mm) O.C. staggered.

NOTE: Refer to the "Corners" detail section of this Stucco Resource Guide.

N. Indicate the type of trim accessory joint (control joint, expansion joint, reveal or other style):

Indicate name of manufacturer (optional):

O. Attach trim accessory joints to building structure, with appropriate fastener selected from list in section 3.38-G. Spacing of fasteners 12 inches (300 mm) O.C. staggered.

NOTE: Refer to "Trim Accessory Joints" detail section of this Stucco Resource Guide.

Joints are recommended if the area is 150 to 180 square feet (14 to 17 m^2). Maximum recommended length of a panel is 18 feet (6 m). Panel size should not exceed a 3-to-l ratio.

The location of trim accessory joints and the type are by the architect.

The design layout of the trim accessory joints and the location of framing members may require the control joints to be attached to the sheathing and/or attached with tie wire to the lath.

The application of sealants is recommended in conjunction with the installation of trim accessory joints. Refer to section 3.04 and details in the "Trim Accessory Joints" section of this Stucco Resource Guide.

A continuous (unbroken) weather-resistant barrier is required behind all trim accessory joints.

Control joints are the type of trim accessory joints recommended for installation over continuous (unbroken) lath. Expansion joints and reveals are types of trim accessory joints which require the lath to be broken behind the joint.

If attachment of control joint is to lath or other trim accessories, use 18-gauge wire tie 12 inches (300 mm) O.C. to each flange, staggered.

Aluminum and/or PVC reveals require that when the lath is installed over the flange, it totally covers it. The welded wire and woven wire lath shall be installed so as the crotch of the lath is over the flange.

P. Installation of expansion joints and reveals requires lath to be cut and attached to both sides of the joint flange.

Q. Indicate type of self-furring lath:

Indicate name of manufacturer (optional):

(Refer to section 2.07 for lath selection.)

NOTE: If type of lath is not specified, the contractor installing the lath selects the type for the project. Lath selected by contractor is to be installed only after it has been approved.

Over sheathing board, lath must be self-furring type and or use furring fasteners.

It is recommended that paperback lath not be used on framed sheathed construction. Installing the weather-resistant barrier separate from the lath installation provides a better flashing method and a uniform thickness of the stucco. On vertical open-frame construction, paperback lath or line wire and paper are required.

R. Attach lath to building structure, with appropriate fastener selected from list in section 3.38-G.

NOTE: Refer to detail section of this Stucco Resource Guide.

Install lath with long dimension of sheets/rolls horizontal, perpendicular to framing members.

Install so overlapping areas are uniform and per manufacturer's recommendations and building code requirements.

Install so as lath is level, flat and taut.

Bend lath and continue around corners to next framing member or install corner beads.

S. Attachment of 3.4-pound-per-square-yard diamond mesh lath, woven wire and welded wire fabric lath shall be at 16" (400 mm) O.C.

NOTE: Refer to section 3.36 on trim accessory and lath attachment.

All spacing is maximum.

3.39 APPLICATION OF STUCCO BASECOAT

NOTE: Verify that trim accessories and lath are properly secured.

Verify trim accessories and lath are installed so as to provide proper screeds, thickness and alignment for the plastering operation.

Do not begin plastering work until unsatisfactory conditions are resolved.

Cement plaster can be applied by trowel or machine.

Apply basecoat to entire wall panel and/or section without interruption or "cold joints."

Install stucco basecoat in accordance with the requirements and references in this Guide Specification and the project construction documents.

> A. Indicate the permissible mix number and its proportions to be used for the basecoat:
>
> Refer to Section 2.15.
>
> **NOTE:** Delete basecoat mixes not applicable to this project.
>
> Do not change selected mix or proportions after plastering operation has started.
>
> Measurement of ingredients (materials of the mix) to be done with devices of known volume, accurate, and successive batches proportion-ally alike. Use only enough mix water to make plaster a workable consistency.
>
> Admixtures are for improving the basecoat mix or enhancing the plaster application. Use products manufactured strictly for this purpose and in accordance with manufacturer's recommendations.
>
> B. Apply stucco first ("scratch") coat in a nominal thickness of 3/8 inch (10 mm).
>
> Refer to Plaster Thickness Tables in this Stucco Resource Guide. First coat to completely embed the lath. First coat to be thick enough to go beyond lath so as to allow for scoring of cement plaster surface.
>
> **NOTE:** Scoring of the first coat should be uniform and shallow, 1/8 inch (3 mm).
>
> C. Apply stucco second ("brown") coat in a nominal thickness of 3/8 inch (10 mm) over stucco first coat. Second coat thickness to bring the combined basecoats (first and second) thickness to a nominal thickness of 3/4 inch (19 mm). (Refer to Plaster Thickness Tables).
>
> **NOTE:** Apply second coat over a damp first coat. If required, apply a fine spray of clean water, so as to dampen only. Do not saturate. Allow water sheen to disappear before applying the second coat.
>
> D. Apply the second coat with sufficient material and pressure to ensure a tight uniform bond to the first coat.
>
> E. The "double-back" method of applying successive coats is recommended. This procedure has little or no delay between applying the second coat over the first coat. .
>
> **NOTE:** Advantages of this method are that it creates a better bond between the two coats, it provides for uniform and better curing of the basecoat and it reduces delay on the project.
>
> Apply second coat as soon as the first coat is rigid enough to receive it.

This method is not recommended for open-frame construction (vertical and horizontal surfaces).

F. Rod the second coat to a true, even plane, filling surface defects with cement plaster.

G. Trowel-float the second coat surface uniformly.

NOTE: The floating process densifies the basecoat and provides a proper surface for the finish coat application.

Float the basecoat after it has set and when moisture is still present in it.

(The float should not adhere to the surface that is to be worked)

Floating the basecoat that is to receive an acrylic finish coat is critical because of the thickness of this finish.

3.40   CURING OF BASECOAT

A. Climatic conditions dictate the need for moist-curing.

NOTE: Cement plaster gains strength in the first day or two with proper moist curing.

Moist-curing enables the cement materials to hydrate properly and the stucco membrane to reach its desired physical properties.

Moist-curing helps prevent "craze cracking."

Cement plaster basecoat to cure for a minimum of seven days before starting finish coat application. Moist cure for the first two days.

B. Moist-curing is recommended when the ambient temperature is 77 F (25 C) or higher and/or when the ambient relative humidity is below 70 percent and the conditions are windy.

NOTE: 4. Moist-cure in the morning and/late afternoon for a period of at least two days.

Moist-cure with a fine mist of clean water; do not saturate.

Moist-cure only after the basecoat has set and is hard.

Extreme weather conditions may require plastic sheets to retard evaporation.

C. The stucco basecoat should be protected from freezing for a period of 24 hours after application.

NOTE: Do not moist-cure if basecoat is subject to freezing.

Do not use frozen materials in mix.

Do not apply cement plaster to a surface that is frozen or contains frost.

3.41   STUCCO FINISH COAT

NOTE: "Stucco finish coat" in this specification refers to the cement plaster finish of which there are two types: (1) Job-site stucco finish coat, and (2) manufactured stucco finish coat.

For colored (integral color) stucco finish coat, the use of a manufactured stucco finish is recommended.

It is recommended that the lighter tones of color (pastel colors) be used for mineral stucco finishes.

Stucco finish coat color uniformity cannot be guaranteed because of a variety of uncontrollable factors (suction of basecoat and the application of the finish coat will vary with climatic conditions). Manufactured stucco finish coat will produce the most consistent color.

Time necessary between the completion of basecoat (brown coat) and the application of finish coat will vary with climatic conditions.

It is recommended that the basecoat which is to receive the finish coat cure dry for a minimum of five days after being moist cured (kept damp) for two days. (This does not pertain to section 3.14 "Finish Coat Direct to Concrete")

For job-site finish coat, it is recommended that the coloring agents be from a stucco finish-coat manufacturer.

- A. Stucco basecoat (or concrete surface) is required to be in a proper condition before application of stucco finish coat or acrylic finish coat.

  NOTE: Do not apply finish coat until all irregularities in the basecoat have been addressed.

- B. Apply stucco finish coat to damp cement plaster basecoat.

  NOTE: Dampen the basecoat with a mist of clean water to obtain uniform suction. Do not saturate: there should not be any visible water on the surface when the finish coat is applied.

- C. Apply stucco finish coat in a nominal thickness of 1/8 inch (3 mm). (Refer to Stucco Thickness Tables).

- D. Apply stucco finish with sufficient material and pressure to ensure a tight bond with basecoat (brown coat) or concrete surface.

- E. Apply stucco finish to a uniform thickness and in a consistent finish in accordance with style of finish specified.

  NOTE: Apply finish coat starting from the top of the wall surface and work down.

  Apply finish coat with no interruptions; no cold joints.

  Apply finish coat so there are no "scaffold" lines or joint stains.

- F. Moist-curing of finish coat is not recommended, except in severe climatic conditions (e.g., extreme heat, strong winds and low relative humidity).

  NOTE: Moist-curing of the stucco finish coat can cause discoloration.

- G. Indicate the type of stucco finish coat:

  Job-site mixed stucco finish

  Manufactured stucco finish

  NOTE: Delete the type of finish not applicable

- H. Indicate the style of stucco finish

  (Refer to 3.41-1)

  NOTE: Smooth trowel finish is not recommended when the material is cement plaster.

  Very heavy textures may have to be applied in the basecoat because a nominal stucco finish coat is not thick enough.

  It is recommended that a sample of the finish coat be applied to a wall at the project site if possible.

Provide style and color sample of stucco finish coat for approval before starting the application of finish coat. Delete if not applicable.

The approved sample to be maintained on project site until the scope of stucco work is completed and approved.

Use only enough water in stucco finish coat mix to make it workable.

    I. Styles of stucco finish
1. Sand Float Finish (fine, medium or course)
2. Machine Dash Finish (light/medium/heavy)
3. Knockdown Dash Finish
4. Lace Finish
5. Light Comb Finish
6. English Finish
7. Frieze Finish
8. Spanish Finish.

NOTE: Delete finish styles not applicable.

    J. Indicate the color of stucco finish

3.42   ACRYLIC FINISH COAT

NOTE: The acrylic finish to be 100-percent acrylic polymer base.

Basecoat of cement plaster that is to receive an acrylic finish coat shall have been floated and/or have a stucco sand-finish.

Do not apply finish until all irregularities in the basecoat or concrete surface have been addressed.

    A. Indicate style of acrylic finish:

Indicate color of acrylic finish:

Indicate approved manufacturers: .

NOTE: Refer to manufacturer for different styles and color selection.

    B. Apply acrylic finish over stucco basecoat (brown coat) or concrete surface a minimum of 1/16 inch (2 mm) dry thickness.

NOTE: Acrylic finish to be applied per the manufacturer's recommendations.

Stucco basecoat shall be free of efflorescence. Apply acrylic finish coat only at a minimum ambient temperature of 40 of (4° C). This temperature is recommended for a minimum of 24 hours after application.

Acrylic finish shall maintain a wet edge at all times. The finished surface shall have no scaffold or stain lines.

Protect finished surface from climatic conditions until dry.

It can be difficult to achieve a uniform color using a spray-applied acrylic finish. Therefore, it is recommended that a troweled application of the acrylic finish be applied first.

Do not moist-cure acrylic finish.

    C. Apply acrylic finish coat with sufficient material to uniformly and completely cover the basecoat.

3.43 ADDITIONAL TYPES OF FINISHES

A. Factory-mixed 100-percent acrylic-based elastomeric finish
B. High-quality exterior acrylic paint. Portland Cement
   NOTE: Applied per manufacturer's recommendations over a stucco finish or acrylic finish.
C. High-quality exterior elastomeric coating
   NOTE: Applied over a stucco finish coat or acrylic finish coat.

Source Premix Marbletite, Web Solutions:

FINISH SPECIFICATION OUTLINE

Source: Stucco Manufacturers Association

10. STUCCO FINISHES

10.1 GENERAL REQUIREMENTS

10.1.1 Packaging

Each manufacturer shall pack manufactured stucco in sealed, multi-wall bags bearing his name, brand, weight, and color identification.

10.1.2 Additives

No fire clay, asbestos, or any other material except clean water shall be added to manufactured stucco.

10.2 EXTERIOR STUCCO (SEE NOTE 1)

10.2.1 Uses

Over any properly prepared Portland cement base (plaster, concrete or masonry).

10.2.2 Materials

A packaged blend of Portland cement (ASTM C-150), hydrated lime (C-206), and properly graded quality aggregate, with or without color.

10.2.3 Properties

When tested per ASTM C-I09-63, exterior stucco shall have a minimum compressive strength of 1200 psi.

10.3 FINISH PLASTER BEDDING COAT (SEE NOTE 2)

10.3.1 Uses

As a bedding coat to receive exposed aggregate.

10.3.2 Materials

A packaged blend of Portland cement (ASTM C-150), hydrated lime (C-206), and properly graded quality aggregate, with or without color.

10.3.3 Properties

When tested per ASTM C-109-63, the finish plaster bedding coat shall have a minimum compressive strength of 2000 psi.

10.4 SMOOTH CEMENT FINISH

10.4.1 Uses

Over any properly prepared Portland cement base (plaster, concrete or masonry). In wet rooms and high humidity areas (shower rooms, lavatories, etc.) or in areas subject to extreme abuse (handball courts, etc.).

10.4.2 Materials

A packaged blend of Portland cement (ASTM C-150), hydrated lime (C-206) and properly graded quality aggregate, without color.

10.4.3 Properties

When tested per ASTM C-109-63, smooth cement finish shall have a minimum compressive strength of 2000 psi.

10.4.4 Curing

Except during damp weather, surface shall be dampened slightly 12 hours after completion and re-dampened at intervals until it hardens.

10.5 PORTLAND CEMENT STUCCO PAINT (SEE NOTE 3)

10.5.1 Uses

On porous surfaces of masonry, concrete, stucco, common brick, masonry block and rough plaster as a decorative, protective, and water-repellent coating.

10.5.2 Materials and Properties

Shall conform to Federal Specification TT-P-21.

10.6 ACOUSTIC TYPE FINISH (EXTERIOR) (SEE NOTE 4)

10.6.1 Uses

Over any properly prepared Portland cement base, such as plaster, concrete, or masonry where texture of interior acoustic ceilings must be matched, or where such a texture is desired on exterior horizontal surfaces. Specify only thickness required to achieve desired texture.

10.6.2 Materials

A packaged blend of Portland cement (ASTM C-150), hydrated lime (C-206), and vermiculite aggregate (C-3 5).

10.10 EXPOSED AGGREGATE FINISH PLASTER COAT (MARBLECRETE)

Source: California Lathing & Plastering Contractors Association, Inc.

10.10.1 Description

Finish plaster shall consist of exposed natural or integrally colored aggregate, partially embedded in a natural (or) colored bedding coat of Portland cement plaster.

10.10.2 Location

Apply finish plaster in areas where shown on the drawings, or called for in the finish schedule or in these specifications. (See Note 1.)

10.10.3 Materials

Lathing and plastering materials for finish plaster shall be standard lathing and plastering materials. (See Note 2.)

10.10.31 Aggregate

For finish plaster finish shall consist of marble chips or pebbles, and shall be clean and free from harmful amounts of dust and other foreign matter. (See Note 3.) Size of finish plaster aggregate shall conform to the following grading standard:

CHIP SIZE	PASSING SCREEN	RETAINED ON SCREEN
Number	Inches	Inches
0	1/8	1/16
1	1/4	1/8
2	3/8	¼
3	1/2	3/8
4.	5/8	1/2

NOTE - Chips larger than No. 4 must be applied manually. Use only for random accent. Check with stucco manufacturer for local availability of aggregate desired.

Aggregate shall be blended by sizes in the percentages of graded chips called for in the specified sample.

10.10.32 Liquid Bonding Agent

See 6.10.1. California Reference Specifications. (See Note 4.)

10.10.33 Sealer

Waterproofing shall be a clear penetrating liquid; (or) glaze shall be a clear non-penetrating liquid. Sealer shall be non-staining and shall resist deterioration from weather exposure.

Apply as directed by the manufacturer. (See Notes.)

10.10.4 Bases

Apply finish plaster over (a) concrete, (b) masonry, (c) Portland cement plaster base coat, (d) gypsum plaster basecoat. (See Note 6.)

10.10.41 Concrete & Masonry

Give masonry and poured concrete surfaces which are to receive finish plaster a dash bond coat of Portland cement plaster; or treat with a liquid bonding agent. (See Note 7.)

Apply plaster using one of the following option methods: (See Note 8).

(a) Apply bedding coat over dash bond coat or liquid bonding agent and double back to required thickness.

(b) Apply brown (leveling) coat over dash bond coat or liquid bonding agent and straighten with rod and darby before applying bedding coat.

10.10.42 Basecoat Plaster

Over metal or wire fabric lath apply basecoat plaster in one of the following optional methods:

(a) Apply scratch coat and allow to set. Apply brown (leveling) coat minimum three-eighth inch (3/8 in.) thick and straighten with rod and darby. Leave rough and allow to set before applying bedding coat.

(b) Apply same as above except that bedding coat may be applied as soon as brown coat is firm enough to support bedding coat without sagging, sliding, or otherwise affecting bond.

(c) Apply scratch coat minimum one-half inch (½ in.) thick using double-back method, cover lath completely and straighten with rod and darby. Scratch horizontally and allow to set before applying bedding coat. Minimum overall thickness of scratch and bedding coat shall be one inch (1 in.).

Over gypsum lath apply brown coat minimum three-eighths inch (3/8 in.) thick, straighten with rods and darby, leave rough and allow to set before applying bedding coat.

10.10.5 Bedding Coat

Factory prepared bedding coat shall be a Portland cement and lime plaster meeting "Specifications and Standards for Manufactured Stucco Finishes" as published by the Stucco Manufacturers Association, Inc., 2402 Vista Nobleza, Newport Beach, CA 92660. (See Note 9.)

Job proportioned bedding coat shall be composed of one part Portland cement, one part Type S lime, and maximum three parts of graded white or natural sand by volume. (See Note 10.)

10.10.51 Mixing

Mix manufactured and job proportioned bedding coat with only sufficient water to attain proper consistency for application and embedment of aggregate.

10.10.52. Thickness

Thickness of bedding coat shall be determined by maximum size of aggregate specified and shall conform to the following: (See Note 11).

Bedding Coat Thickness	Aggregate Size (Max.)	
Minimum Inches	Number	Inches
3/8	# 0	1/8
3/8	# 1	1/4
3/8	# 2	3/8
3/8	# 3	1/2
1/2	# 4	5/8
80% of aggregate dimension	Larger than #4	

10.10.6 Application of Finish plaster

Apply bedding coat to proper thickness and straighten to a true, reasonably smooth surface with rod and darby.

Allow bedding coat to take up until it attains the proper consistency to permit application of aggregate.

Apply the aggregate to the bedding coat, starting at the perimeter of a panel area and working towards the center.

Tamp lightly and evenly to assure embedment of the aggregate and to bring surface to an even plane.

10.10.7 Curing

Portland cement finish plaster shall retain sufficient moisture for hydration (hardening) for 24 hours minimum. Where weather conditions require, keep finish plaster damp by spraying lightly.

10.10.8 Samples

Furnish samples in duplicate (triplicate). Samples shall conform to the following:

- A. Screed and Casing Thickness — Inch
- B. Bonding Agent
- C. (1) Bedding Coat Thickness — Inch
   (2) Bedding Coat Color Number — Number, by Mfr.
   (3) Aggregate Sizes & Percentages — Size   Pct.
   No.1   %
   No.2   %
   No.3   %
   No.4   %

   (4) Aggregate
   Description:
   Color:
   Source:
   Producer:
- D. Finish plaster
   Number:
   by (Manufacturer):
- E. Sealer
   Name:
   by (Manufacturer):

10.10.9 Murals

The architect or his representative shall transfer design pattern shown on large scale drawings to base which receives finish plaster mural. The transferred design shall serve as guide for the application of flexible or rigid metal separating screeds. Artist's rendering, and color and aggregate key, shall be provided contractor at time of bidding.

NOTES

This specification reference presupposes that basic lathing and plastering specifications have already been prepared for the project and relates only to materials and application methods for finish plaster.

NOTE 1

Specify in lathing section control joints, casing beads, parting screeds, or other metal sections to define panels as shown on drawings, and to serve as grounds for finish plaster. Finish plaster panels should not exceed 120 square feet, or 11 feet in any direction. Specify total ground thickness and show on drawings.

NOTE 2

E.g. Portland Cement; Lime; Sand; Gypsum, etc.

NOTE 3

Other natural aggregate such as quartz, cinders, sea shells, or integrally colored manufactured aggregate, such as crushed glass, china, ceramics, may be specified providing they are weather resistant, permanent in color, moderately hard (3 or more on the MOH scale) and are compatible with the bedding coat.

NOTE 4

Liquid bonding agent may be applied to plaster basecoat before application of bedding coat. If bonding agent is desired, so specify; or permit use at option of contractor.

NOTE 5

Sealer is recommended to improve and retain color and cleanliness of finish plaster.

NOTE 6

Specify those bases applicable to project.

NOTE 7

See 7.1.2 Masonry Surfaces; 7.1.3 Poured Concrete Surfaces, Calif. Reference Specifications.

NOTE 8

Check with local building officials for method of application approved. All building codes require minimum overall thickness of exterior Portland cement plaster to be 7/8 inch; gypsum plaster minimum 1/2 inch.

NOTE 9

See 10.3. Specify pertinent information under Samples, 10.10.8.

NOTE 11

Overall thickness will depend on thickness of basecoat plaster plus thickness of bedding coat.

APPENDIX D

Scaffolding and Safety

SAFETY GUIDELINES

Serious injury or death can result from your failure to familiarize yourself, and comply with all applicable safety requirements of federal, state, and local regulations and these safety guidelines before erecting, using, or dismantling this scaffold.

Figure D-1: Scaffolding in place

ERECTION OF SCAFFOLDING

PRIOR TO ERECTION – ALL SCAFFOLD ASSEMBLIES

- You should inspect jobsite to determine ground conditions or strength of supporting structure, and for proximity of electric power lines, overhead obstructions, and wind conditions, the need for overhead protection or weather protection coverings. These conditions must be evaluated and adequately provided for.
- Frame spacing and mudsill size can only be determined after the total loads to be imposed on the scaffold and the strength of the supporting soil or structure are calculated and considered. A qualified person must do this analysis. Load carrying information on components is available from the manufacturer.
- Stationary scaffolds over 125 feet in height and rolling scaffolds over 60 feet in height must be designed by a professional engineer.
- All equipment must be inspected to see that it is in good condition and is serviceable.
- Damaged or deteriorated equipment should not be used.
- Wood plank should be inspected to see that it is graded for scaffold use, is sound, and in good condition, straight grained, free from saw cuts, splits, and holes. (Not all species and grades of lumber can be used as scaffold plank. Wood planks used for scaffolding must be specifically graded for scaffold use by an approved grading agency).

- The scaffold assembly must be designed to comply with local, State, and Federal safety requirements.

ERECTION OF FIXED SCAFFOLD

- Scaffold must be erected, moved, or disassembled only under the supervision of qualified persons. All persons erecting, moving, dismantling, or using scaffolding must wear hard hats.
- Mudsills must be adequate size to distribute the loads on the scaffolding to the soil or supporting structure. Special care is needed when scaffolding is to be erected on fill or other soft ground or on frozen ground. Sills should be level and in full contact with the supporting surface.
- Base plates or screw jacks, with base plates must be in firm contact with both the sills and the legs of the scaffolding. Compensate for uneven ground with screwjacks with base plates.
- DO NOT USE unstable objects such as blocks, loose bricks, etc.
- Plumb and level scaffold until connections can be made with ease. Do not force members to fit. Be sure scaffold stays level and plumb as erection progresses
- Ties, guys, bracing and/or outriggers may be needed to assure a safe stable scaffold assembly. The height of the scaffold in relation to the minimum base width, wind loads, the use of brackets or cantilevered platforms and imposed scaffold loads determines the need for stability bracing. The following general guides are minimum requirements.
- Federal OSHA requires that scaffolding must always be secure when the height of the scaffold exceeds for (4) times the minimum base width. (California requires stability bracing when the scaffold height exceeds three (3) times the minimum base width).
- The bottom tie must be placed no higher than four (4) times the minimum base width and every 26 feet vertically thereafter. Ties should be placed as close to the top of the scaffold as possible and, in no case, less than four (4) times (three (3) times in California) the minimum base width of the scaffold from the top.
- Vertical ties should be placed at the ends of scaffold runs and at no more than 30 feet horizontal intervals in between.
- Ties should be installed as the erection progresses and not removed until the scaffold is dismantled to that height.
- Side brackets, cantilevered platforms, pulleys or hoist arms and wind conditions introduce overturning and uplift forces that must be considered and compensated for. These assemblies may require additional bracing, tying, or guying.
- Circular scaffolds erected completely around or within a structure may be restrained from tipping by the use of "stand off" bracing members.
- Each leg of a freestanding tower must be guyed at the intervals outlined above or otherwise restrained to prevent tipping or overturning.
- Work platforms must be fully planked either with scaffold graded solid sawn or laminated plank, in good sound condition, or with fabricated platforms in good condition.
- Each plank must overlap the support by a minimum of 6 inches or be cleated, i.e. 8-foot planks on 7-foot spans must be cleated.
- Plank should not extend beyond the support by more than 18 inches. Such overhangs should be separated from the work platform by guard railing so that they cannot be walked on.

- Plank on continuous runs must extend over the supports and overlap each other by at least 12 inches.
- Spans of full thickness, 2 inch by 10-inch scaffold grade planks, should never exceed 10 feet. Loads on plank should be evenly distributed and not exceed the allowable loads for the type of plank being used. No more than one person should stand on an individual plank at one time.
- Planks and/or platforms should be secured to scaffolding when necessary to prevent uplift of displacement because of high winds or other job conditions.
- Guardrails must be used on all open sides and ends of scaffold platforms. Both top and midrails are required. Local codes specify the minimum heights where guardrails are required, however, use at lower heights if falls can cause injury.
- Toe boards are required whenever people are required to work or pass under or around the scaffold platform.
- Access must be provided to all work platforms. If it is not available from the structure, access ladders, frames with built-in ladders, or stairways must be provided. When frames with built-in ladders are used, cleated plank or fabricated plank must be used at platform levels to minimize or eliminate platform overhang. Access ladders must extend at least three (3) feet above platforms.
- Side and end brackets are designed to support people only. Materials should never be placed on cantilevered platforms unless the assembly has been designed to support material loads by a qualified person. (These types of platforms cause overturning and uplift forces, which must be compensated for. All frames should be fastened together to prevent uplift an overturning moment compensated for with counterweights or adequate ties).
- Putlogs must never be used for the storage of materials. They are designed for personnel use only. Special care should be taken when putlogs are used.
- Putlogs should overhang the support points by at least 6 inches. Use putlogs hangers with bolts fastened to support putlogs on frames.
- Putlog spans of greater than 12 feet require kneebracing and lateral support.
- Putlogs used as side or end brackets need special bracing.
- Bridging between towers should not be done with plank or stages unless the assembly is designed by a qualified person and overturning moments have been compensated for.
- Scaffold should not be used as material hoist towers or for mounting derricks unless a qualified person designs the assembly.
- Check the erected assembly before use. A qualified person should thoroughly inspect the completed assembly to see that is complies with all safety codes, that nuts and bolts are tightened, that it is level and plumb, that work platforms are fully planked, that guardrails are in place and safe access is provided.

ERECTION OF ROLLING SCAFFOLDS

- Height of the tower must not exceed four (4) times the minimum base dimension (three (3) times in California). Outrigger frames or outrigger units on both sides of the tower may be used to increase base width dimension when necessary.
- All casters must be secured to frame legs or screwjacks with a nut and bolt or other secure means. Total weight of tower should not exceed the capacity of the casters.
- Screw jacks must not be extended more than 12 inches above caster base. Tower must be kept level and plumb at all times.

- Horizontal/diagonal bracing must be used at the bottom and top of tower and at intermediate levels of 20 feet. Fabricated planks with hooks may replace the top diagonal brace.
- All frames must be fully cross-braced.
- Only prefabricated plank or cleated plank should be used.
- Casters must be locked at all times the scaffold is not being moved.

USE OF SCAFFOLDS

ALL SCAFFOLDS

- Inspect the scaffold assembly before each use to see that it is assembled correctly, that it is level and plumb, base plates are in firm contact with sills, bracing is in place and connected, platforms are fully planked, guardrails in place, safe access is provided, that it is properly tied and/or guyed and that there are no overhead obstructions or electric lines within 12 feet of the scaffold assembly.
- Use only the safe means of access that is provided. Do not climb bracing or frames not specifically designed for climbing. If such access is not provided, insist that it be provided.
- Climb Safely
    - Face the rungs as you climb up or down.
    - Use both hands.
    - Do not try to carry materials while you climb.
    - Be sure of your footing and balance before you let go with your hands. Keep one hand firmly on frame or ladder at all times.
    - Do not work on slippery rungs to avoid slipping.
- Do not overload platforms with materials.
- Working heights should not be extended by planking guardrails or by use of boxes or ladders on scaffold platforms.
- Do not remove any component of a completed scaffold assembly except under the supervision of a qualified person. Any component that has been removed should be immediately replaced.

ROLLING TOWERS

All of the above precautions plus:
- Do not ride manually propelled rolling scaffold. No personnel should be on the tower while it is being moved.
- Lock all casters before getting on the tower.
- Work only within the platform area: do not try to extend overhead work area by reaching out over guardrailing.
- Do not bridge between two rolling towers with plank or stages.
- Secure all materials before moving scaffolds.
- Be sure floor surface is clear of obstructions or holes before moving scaffold.
- Be sure there are no overhead obstructions or electric power lines in the path of rolling scaffold.
- Rolling towers must only be used on level surfaces.

■ Move rolling towers by pushing at the base level only. Do not pull from the top.

PART 1910

The following are excerpts from the Manual of Accident Prevention in Construction, U.S. Department of Labor, OSHA.

1910.28 SAFETY REQUIREMENTS FOR SCAFFOLDING

(a) General Requirements for All Scaffolds

(1) Scaffolds shall be furnished and erected in accordance with this standard for persons engaged in work that cannot be done safely from the ground or from solid construction, except that ladders used for such work shall conform to § 1910.25 and § 1910.26.

(2) The footing or anchorage for scaffolds shall be sound, rigid, and capable of carrying the maximum intended load without settling or displacement. Unstable objects such as barrels, boxes, loose brick, or concrete blocks shall not be used to support scaffolds or planks.

(3) Guardrails and toeboards shall be installed on all open sides and ends of platforms more than 10 feet above the ground or floor except:

(i) Scaffolding wholly within the interior of a building and covering the entire floor area of any room therein and not having any side exposed to a hoistway, elevator shaft, stairwell, or other floor openings, and

(ii) Needle-beam scaffolds and floats in use by structural iron workers.

Guardrails should all be 2 x 4 inches or the equivalent, installed no less than 36 inches or not more than 42 inches high, with a midrail, when required, of 1- x 4-inch lumber or equivalent. Supports should be at intervals not to exceed ten feet. Toeboards shall be a minimum of 4 inches in height.

(4) Scaffolds and their components shall be capable of supporting without failure at least four times the maximum intended load.

(5) Scaffolds and other devices mentioned or described in this section shall be maintained in safe condition. Scaffolds shall not be altered or moved horizontally while they are in use or occupied.

(6) Any scaffold damaged or weakened from any cause shall be immediately repaired and shall not be used until repairs have been completed.

(7) Scaffolds shall not be loaded in excess of the working load for which they are intended.

(8) All load-carrying timber members of scaffold framing shall be a minimum of 1,500 f. (Stress Grade) construction grade lumber. All dimensions are nominal sizes as provided in the American Lumber Standards, except that where rough sizes are noted, only rough or undressed lumber of the size specified will satisfy minimum requirements. (NOTE: Where nominal sizes of lumber are used in place of rough sizes, the nominal size lumber shall be such as to provide equivalent strength to that specified in tables D-7 through D-12 and D-16.)

(9) All planking shall be Scaffold Grade as recognized by grading rules for the species of wood used. The maximum permissible spans for 2- x 9-inch or wider planks are shown in the following table:

MATERIAL					
	Full Thickness Undressed Lumber			Nominal Thickness Lumber	
Working Load (p.s.f.)	25	50	75	25	50
Permissible Span (ft.)	10	8	6	8	6

The maximum permissible span for 1¼ x 9-inch or wider plank of full thickness is 4 feet with medium loading of 50 p.s.f.

(10) Nails or bolts used in the construction of scaffolds shall be of adequate size and in sufficient numbers at each connection to develop the designed strength of the scaffold. Nails shall not be subjected to a straight pull and shall be driven full length.

(11) All planking or platforms shall be overlapped (minimum 12 inches) or secured from movement.

(12) An access ladder or equivalent safe access shall be provided.

(13) Scaffold planks shall extend over their end supports not less than 6 inches nor more than 18 inches.

(14) The poles, legs, or uprights of scaffolds shall be plumb, and securely and rigidly braced to prevent swaying and displacement.

(15) Materials being hoisted onto a scaffold shall have a tag line.

(16) Overhead protection shall be provided for men on a scaffold exposed to overhead hazards.

(17) Scaffolds shall be provided with a screen between the toeboard and the guardrail, extending along the entire opening, consisting of No. 18 gauge U.S. Standard Wire one-half-inch mesh or the equivalent, where persons are required to work or pass under the scaffolds.

(18) Employees shall not work on scaffolds during storms or high winds.

(19) Employees shall not work on scaffolds which are covered with ice or snow, unless all ice or snow is removed and planking sanded to prevent slipping.

(20) Tools, materials, and debris shall not be allowed to accumulate in quantities to cause a hazard.

(21) Only treated or protected fiber rope shall be used for or near any work involving the use of corrosive substances or chemicals.

(22) Wire or fiber rope used for scaffold suspension shall be capable of supporting at least six times the intended load.

(23) When acid solutions are used for cleaning buildings over 50 feet in height, wire rope supported scaffolds shall be used.

(24) The use of shore scaffolds or lean-to-scaffolds is prohibited.

(25) Lumber sizes, when used in this section, refer to nominal sizes except where otherwise stated.

(26) Scaffolds shall be secured to permanent structures, through use of anchor bolts, reveal bolts, or other equivalent means. Window cleaners' anchor bolts shall not be used.

(27) Special precautions shall be taken to protect scaffold members, including any wire or fiber ropes, when using a heat-producing process.

(b) General Requirements for Wood Pole Scaffolds
  (1) Scaffold poles shall bear on a foundation of sufficient size and strength to spread the load from the poles over a sufficient area to prevent settlement. All poles shall be set plumb.
  (2) Where wood poles are spliced, the ends shall be squared and the upper section shall rest squarely on the lower section. Wood splice plates shall be provided on at least two adjacent sides and shall not be less than 4 feet 0 inches in length, overlapping the abutted ends equally, and have the same width and not less than the cross-sectional area of the pole. Splice plates of other materials of equivalent strength may be used.
  (3) Independent pole scaffolds shall be set as near to the wall of the building as practicable.
  (4) All pole scaffolds shall be securely guyed or tied to the building or structure. Where the height or length exceeds 25 feet, the scaffold shall be secured at intervals not greater than 25 feet vertically and horizontally.
  (5) Putlogs or bearers shall be set with their greater dimensions vertical, long enough to project over the ledgers of the inner and outer rows of poles at least 3 inches for proper support.
  (6) Every wooden putlog on single pole scaffolds shall be reinforced with a 3/16- x 2-inch steel strip or equivalent secured to its lower edge throughout its entire length.
  (7) Ledgers shall be long enough to extend over two pole spaces. Ledgers shall not be spliced between the poles. Ledgers shall be reinforced by bearing blocks securely nailed to the side of the pole to form a support for the ledger.
  (8) Diagonal bracing shall be provided to prevent the poles from moving in a direction parallel with the wall of the building, or from buckling.
  (9) Cross bracing shall be provided between the inner and outer sets of poles in independent pole scaffolds. The free ends of pole 'scaffolds shall be cross braced.
  (10) Full diagonal face bracing shall be erected across the entire face of pole scaffolds in both directions. The braces shall be spliced at the poles.
  (11) Platform planks shall be laid with their edges close together so the platform will be tight with no spaces through which tools or fragments of material can fall.
  (12) Where planking is lapped, each plank shall lap its end supports at least 12 inches. Where the ends of planks abut each other to form a flush floor, the butt joint shall be at the centerline of a pole. The abutted ends shall rest on separate bearers. Intermediate beams shall be provided where necessary to prevent dislodgment of planks due to deflection, and the ends shall be nailed or cleated to prevent their dislodgement.
  (13) When a scaffold turns a corner, the platform planks shall be laid to prevent tipping. The planks that meet the corner putlog at an angle shall be laid first, extending over the diagonally placed putlog far enough to have a good safe bearing, but not far enough to involve any danger from tipping. The planking running in the opposite direction at right angles shall be laid so as to extend over and rest on the first layer of planking.

SCAFFOLDING SAFETY

14-1 GENERAL

Scaffolds and other elevated work platforms are responsible for many accidents through falls and falling objects. Properly designed and constructed scaffolding and staging may be used with no greater hazard than any other work area; improper or makeshift scaffolds, on the other hand, generally prove to be sources of trouble.

Scaffolds should be designed, built, and inspected by competent persons. To avoid the use of makeshift platforms, each job should be carefully examined ahead of time so that all necessary ramps and platforms can be provided when needed.

14-2 TUBULAR STEEL SCAFFOLDING

14-2.1 General

Steel scaffolds are more durable than wood structures, and reduce fire exposure. Steel scaffolding should be erected and used in accordance with manufacturer's recommendations.

Proper seating and locking of all connections with the correct devices is of extreme importance.

14-2.2 Footings

Firm footing must be provided for each upright; a metal plate is most satisfactory and may be provided with scaffolding. It is necessary to supplement this plate with planking or other support in loose material; minimum thickness of lumber recommended for this purpose is two inches. Footings should be secured against movement by recessing, staking, or other means.

14-2.3 Uprights

All uprights must be plumb. For scaffolds less than 75 feet high, a minimum outside diameter of 2 inches is recommended for tubing. For scaffolding above this height, the uprights should be in accordance with manufacturer's recommendations.

14-2.4 Platforms

Platforms should be constructed of 2 x 10 wood planks. As most ledgers will be single members, platform planking will usually be lapped for continuity. All lap joints should be made at ledgers, with a minimum six inch lap on each side of ledger member.

14-2.5 Toeboards

Toeboards are usually nailed to uprights when erecting wood pole scaffolds. Because tubular scaffold uprights are metal. Toe boards must be nailed to platform planks or fastened to uprights with bolts or other suitable connections.

14-2.6 Guard Rails

Guard rails must be secured to uprights by connectors designed for this use. Guard rails for tubular scaffolds should have tubing with a minimum outside diameter of 1½ inches.

The most frequent abuse of safe practice in the use of the tubular scaffolds is the omission of toeboards and guardrails.

Including these items adds very little to the cost and time of scaffold erection and will be paid for in terms of safe working conditions.

14-2.7 Bracing

Tubular scaffold uprights are generally much smaller in diameter than timber posts for the same size scaffold. It is extremely important that:

(1) uprights be erected and maintained in vertical (plumb) position. and
(2) diagonal bracing be provided.

Exterior scaffolds should be tied or anchored to the building at a height of 3 times the narrowest width and every two sections thereafter as a minimum. As work progresses upward and platforms are removed. It is important that all ledgers be left in place to provide rigidity. If scaffold is tied into masonry construction, all ledgers should be securely set into vertical masonry joints by means of a suitable clincher plate with a minimum thickness of 3/16 inch.

14-2.B Rolling Scaffolds

Workmen should not ride rolling scaffolds or attempt to move rolling scaffolds by pulling on overhead pipes or structures. All material and equipment should be removed from platform or secured before moving scaffold. When moving rolling scaffolds, watch out for holes and overhead obstructions. Caster brakes should be applied at all times when scaffolds are not being moved.

Adjusting screws should not be extended more than 12". Use horizontal diagonal bracing near top, bottom, and at intermediate levels of 30 feet. The working platform height of a rolling scaffold should not exceed four times the smallest base dimension unless guyed or otherwise stabilized. Overturning effect should be considered when using brackets on rolling scaffolds.

14-3 WOOD SCAFFOLDING

14-3.1 Lumber

All lumber used in constructing ramps, platforms, staging, scaffolding, etc., should be of good quality, seasoned, and straight-grained, free of large loose or dead knots, knots in groups, checks, splits, and other defects which tend to decrease the structural strength.

14-3.2 Nails

All nails should be driven home. No nail should be subjected to direct pull. A minimum of four nails per joint is recommended. The size of the nail used will depend upon the load that must be carried by the joint and the thickness of the material being jointed: one-inch stock requiring 8d nails, two-inch stock requiring 16d nails, etc.

Double-headed nails are recommended if structure is to be dismantled for re-use.

14-3.3 Members and Connections

Structural members, fasteners, handrails, etc. should be clean and rust-free. Maintenance of a sound paint coat on metal parts will assist in this task. Patent connections, ladder-jacks, lifting devices, and other hardware should be kept in working order and maintained in accordance with manufacturer's instructions.

14-4 DESIGN

Each scaffold should be designed for the loads, which will be carried on in the performance of the work at hand. All loads, including workmen, building materials, and the weight of the scaffold structure itself must be taken into account. The structure should be designed for a FACTOR OF SAFETY of at least 4.

Adequate footings, such as planks, should be provided for uprights, especially when they rest on earth, sand, or loose material. Cross-bracing to provide stability must be provided. Permanent ladders or stairs should be provided. If ladder is used, it should be secured firmly against slipping and overturning. Ladder may be of standard manufacture or built on the job.

Adequate head protection must be provided for men working on the scaffold if work is being carried out overhead. A roof of lumber, heavy canvas, or screen wire should be used, as appropriate.

Handrails and intermediate rails should be provided on all open sides of working platforms. Screening is recommended. Toeboards should be installed on all open sides of working platforms.

14-5 POLE SCAFFOLDS

14-5.1 Independent Pole Scaffolds

Very little support is derived from the building under construction, however, connections to the building do provide important stability to the scaffold structure and should be made at convenient points.

For purposes of design recommendations, pole scaffolds have been classified "light trades" and "heavy trades." The former includes carpenters, painters, and other trades, which will not bring heavy material loads on the working platform. The "heavy trades" include bricklayers, stonemasons, concrete workers, and steel workers.

14-5.2 Single Pole Scaffolds

Single pole scaffolds differ from independent scaffolds in that only one side is supported by uprights, one end of each ledger being carried by the building under construction itself.

Construction principles are much the same, except that heavier ledgers are recommended, and more numerous connections should be made to insure against the scaffold swinging away from the building.

Ledgers should be notched from 4½ to 5 inches when inserted in a brick wall. If patent steel seating attachments are utilized, care must be taken to insure that the end is securely seated. If spring stays are used to tie the scaffold to the building, place wood block, brick, or other separator near the end of the stay closest to building. Notch wood ledgers from the upper side.

If working platforms are removed as work progresses upward, ledgers should be left in place for structural strength and rigidity. Single pole scaffolds should be cross-braced in both directions, along the face of the building as illustrated for independent scaffolds, and at right angles to the building face at every third or fourth upright.

14-6 HOUSEKEEPING AND INSPECTION

The following general rules are prescribed for maintaining all types of scaffolding in safe working condition.

1. All scaffold structures should be inspected at least daily by the project manager, project engineer, or other responsible person designated by the job superintendent to perform this task.
2. No change of any kind should be made in scaffolds without engineering approval.
3. The structure should be cleared of all rubbish daily. No tools should be left on scaffolds overnight.
4. No excess materials should be stockpiled on scaffolding.

5. Notices regarding the use of scaffolds, when needed, should be conspicuously displayed and observed.
6. Scaffold structures should be protected from trucks and other vehicles, which might come into contact with them.
7. Working platforms should be free of ice, snow, oil, etc., before being used.
8. No open fire should be permitted upon or near scaffolds.

14-7 OUTRIGGER SCAFFOLDS

14-7.1 General

An outrigger scaffold is a working platform supported by cantilevered or braced outrigger beams or "thrustouts."

The use of this type of scaffold is not recommended where any other type can be used. Construction should be assigned only to experienced mechanics.

14-7.2 Thrustouts

Use only first-grade, straight-grained timber, minimum dimension 3 x 10 inches, set on edge. The inner end must be securely and rigidly anchored to the structure. Weights to hold down ends of thrustouts must not be used. Thrustouts must be secured against overturning by the use of blocking or bracing. Steel beams may be used for support-recommended minimum size, 6" I-beams, 12.25 pounds per foot. Maximum recommended extension of thrustouts beyond face of building is 6 feet. Thrustouts should not be built into the wall of the structure and left with no other support. When loading conditions warrant added support, braces extending from the outer ends of thrustouts and secured to the structure below may be utilized.

A stop bolt at the outer end of thrustouts will prevent hanger from working free of beam ends.

It is further recommended that thrustouts be spaced not more than 6 feet center-to-center along the length of the platform.

14-7.3 Platform

Two 2 x 10 planks are the minimum for a platform. Planks should be closely laid and securely nailed to thrustouts. Planks should be lapped over thrustouts a minimum of 12 inches.

The minimum overhang beyond thrustout at platform ends is 4 inches, with a maximum of 12 inches, ends securely nailed and guard-railed.

14-7.4 Toeboards

Toeboards should be provided along the outside edge of platform and at open ends, minimum 1 x 6 recommended.

14-7.5 Railing Uprights

Upright members should be long enough to support top guardrail approximately 42 inches above the platform, and to extend 12 inches below the outrigger. The lower end should be securely braced to the outrigger. Minimum size recommended for uprights is 2 x 4.

Source: Manual of Accident Prevention in Construction by the Associated General Contractors of America, Inc.

Table D-1 / Minimum Nominal Size and Maximum Spacing of Members of Single Pole Scaffolds / LIGHT DUTY

	Maximum height of scaffold	
	20 ft.	60 ft.
Uniformly distributed load	Not to exceed 25 pounds per square foot	
Poles or uprights	2 x 4 in	4 x 4 in.
Pole spacing (longitudinal)	6 ft. 0 in	10 ft. 0 in.
Maximum width of scaffold	5 ft. 0 in	5 ft. 0 in.
Bearers or putlogs to 3 ft. 0 in. width	2 x 4 in	2 x 4 in.
Bearers or putlogs to 5 ft. 0 in. width	2 x 6 in. or 3 x 4 in	2 x 6 in. or 3 x 4 in. (rough)
Ledgers	1 x 4 in	1¼ x 9 in.
Planking	1¼ x 9 in. (rough)	2 x 9 in.
Vertical spacing of horizontal members	7 ft. 0 in	7 ft. 0 in.
Bracing, horizontal and diagonal	1 x 4 in	1 x 4 in.
Tie-ins	1 x 4 in	1 x 4 in.
Toeboards	4 in. high (minimum)	4 in. high (minimum)
Guardrail	2 x 4 in	2 x 4 in.
All members except planking are used on edge.		

Table D-2 / Minimum Nominal Size and Maximum Spacing of Members of Single Pole Scaffolds / MEDIUM DUTY

Uniformly distributed load	Not to exceed 50 pounds per square foot
Maximum height of scaffold	60 ft.
Poles or uprights	4 x 4 in.
Pole spacing (longitudinal)	8 ft. 0 in.
Maximum width of scaffold	5 ft 0 in.
Bearers or putlogs	2 x 9 in. or 3 x 4 in.
Spacing of bearers or putlogs	8 ft. 0 in.
Ledgers	2 x 9 in.
Vertical spacing of horizontal members	9 ft. 0 in.
Bracing, horizontal	1 x 6 in. or 1¼ x 4 in.
Bracing, diagonal	1 x 4 in.
Tie-ins	1 x 4 in.
Planking	2 x 9 in.
Toeboards	4 in. high (minimum)
Guardrail	2 x 4 in.
All members except planking are used on edge.	

Table D-3 / Minimum Nominal Size and Maximum Spacing of Members of Single Pole Scaffolds / HEAVY DUTY

Uniformly distributed load	Not to exceed 75 pounds per square foot
Maximum height of scaffold	60 ft.
Poles or uprights	4 x 4 in.
Pole spacing (longitudinal)	6 ft. 0 in.
Maximum width of scaffold	5 ft. 0 in.
Bearers or putlogs	2 x 9 in. or 3 x 5 in. (rough)
Spacing of bearers or putlogs	6 ft. 0 in.
Ledgers	2 x 9 in.
Vertical spacing of horizontal members	6 ft. 6 in.
Bracing, horizontal and diagonal	2 x 4 in.
Tie-ins	1 x 4 in.
Planking	2 x 9 in.
Toeboards	4-in. high (minimum)
Guardrail	2 x 4 in.
All members except planking are used on	

Table D-4 / Minimum Nominal Size and Maximum Spacing of Members of Independent Pole Scaffolds / LIGHT DUTY

	Maximum height of scaffold	
	20 ft.	60 ft.
Uniformly distributed load	Not to exceed 25 pounds per square foot	
Poles or uprights	2 x 4 in	4 x 4 in.
Pole spacing (longitudinal)	6 ft. 0 in	10 ft. 0 in.
Pole spacing (transverse)	6 ft. 0 in	10 ft. 0 in.
Ledgers	1¼ x 4 in	1¼ x 9 in.
Bearers to 3 ft. 0 in. span	2 x 4 in	2 x 4 in.
Bearers to 10 ft. 0 in. span	2 x 6 in. or 3 x 4 in.	2 x 9 in. (rough) or 3 x 8 in.
Planking	1¼ x 9 in	2 x 9 in.
Vertical spacing of horizontal members	7 ft. 0 in	7 ft. 0 in.
Bracing, horizontal and diagonal	1 x 4 in	1 x 4 in.
Tie-ins	1 x 4 in	1 x 4 in.
Toeboards	4 in. high	4 in. high (minimum)
Guardrail	2 x 4 in	2 x 4 in.
All members except planking are used on edge.		

Table D-5 / Spacing Minimum Nominal Size and Maximum Spacing of Members of Independent Pole Scaffolds / MEDIUM DUTY

Uniformly distributed load	Not to exceed 50 pounds per square foot
Maximum height of scaffold	60 ft.
Poles or uprights	4 x 4 in.
Pole spacing (longitudinal)	8 it. 0 in.
Pole spacing (transverse)	8 ft. 0 in.
Ledgers	2 x 9 in.
Vertical spacing of horizontal members	6 ft. 0 in.
Spacing of bearers	8 ft. 0 in.
Bearers	2 x 9 in. (rough) or 2 x 10 in.
Bracing, horizontal	1 x 6 in. or 1¼ x 4 in.
Bracing, diagonal	1 x 4 in.
Tie-ins	1 x 4 in.
Planking	2 x 9 in.
Toeboards	4 in. high (minimum)
Guardrail	2 x 4 in.
All members except planking are used on edge.	

Table D-6 / Minimum Nominal Size and Maximum Spacing of Members of Independent Pole Scaffolds / HEAVY DUTY

Uniformly distributed load	Not to exceed 75 pounds per square foot
Maximum height of scaffold	60 ft.
Poles or uprights	4 x 4 in.
Pole spacing (longitudinal)	6 ft. 0 in.
Pole spacing (transverse)	8 ft. 0 in.
Ledgers	2 x 9 in.
Vertical spacing of horizontal members	4 ft. 6 in.
Bearers	2 x 9 in. (rough)
Bracing, horizontal and diagonal	2 x 4 in.
Tie-ins	1 x 4 in.
Planking	2 x 9 in.
Toeboards	4 in. high (minimum)
Guardrail	2 x 4 in.
All members except planking are used on edge.	

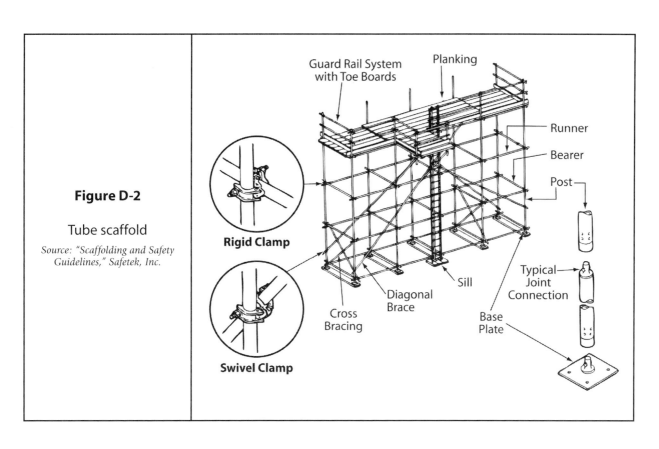

Figure D-2

Tube scaffold

Source: "Scaffolding and Safety Guidelines," Safetek, Inc.

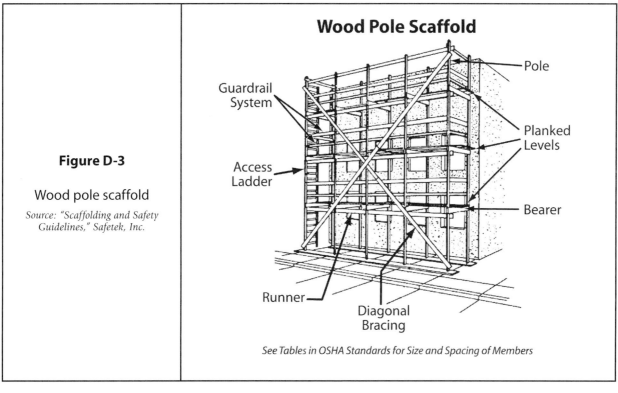

Figure D-3

Wood pole scaffold

Source: "Scaffolding and Safety Guidelines," Safetek, Inc.

245

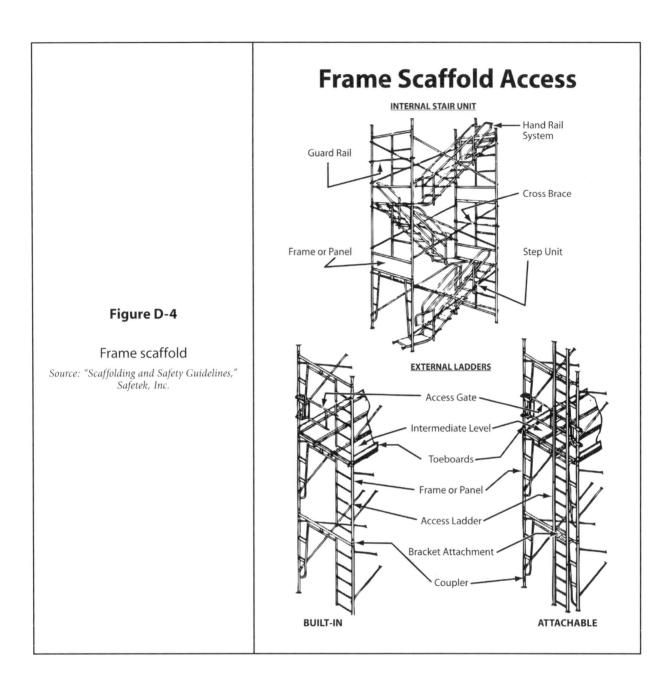

Figure D-4

Frame scaffold

Source: "Scaffolding and Safety Guidelines," Safetek, Inc.

APPENDIX E

Stucco and Related Details

The details appearing on the following pages have been reproduced from the *Stucco Resource Guide – Third Edition* (2002), with the kind permission of its publisher, the Northwest Wall & Ceiling Bureau. For many more details and further research, we highly recommend the original reference, which is available from the publisher at (206) 524-4243 or online at nwcb.org.

FIGURE NO.	DESCRIPTION
E-1	"Open Stud" Construction
E-2	Sheathed Wood Framing Construction
E-3	Sheathed Steel Framing Construction
E-4	Stucco Direct to Concrete Masonry
E-5	Stucco System Attached to Concrete Masonry
E-6	Finish "Skim" Coat Direct to Concrete
E-7	Two Coat Stucco Direct to Concrete
E-8	Stucco System Attached to Concrete
E-9	Stucco Insulation System to Concrete Masonry
E-10	Security Wall
E-11	Opened Framed Rainscreen Construction
E-12	Semi-Rigid Rainscreen Construction
E-13	Drainage Medium Construction
E-14	Sheathed Rainscreen Construction
E-15	Flashing Membrane and Water Resistant Barrier Application Sequence (Steps 1 and 2)
E-16	Flashing Membrane and Water Resistant Barrier Application Sequence (Steps 3 and 4)
E-17	Flashing Membrane and Water Resistant Barrier Application Sequence (Water Resistant Barrier layers)
E-18	Round Window Head Application (Steps 1 and 2)
E-19	Round Window Head Application (Steps 3 and 4)
E-20	Flashing Wrapping Rough Inside Opening Application
E-21	Field Installed Sill Pan Flashing
E-22	Exploded Window Elevation
E-23	Window Head Flashing Assembly
E-24	Casing Bead at Concrete Foundation
E-25	Weep Screed at Concrete Foundation
E-26	Wood Trim at Concrete Foundation
E-27	Termination at Slab/Sidewalk
E-28	Termination at Cantilevered Wall
E-29	Termination at Foundation
E-30	Termination at Foundation/Finished Grade
E-31	Termination at Foundation/Finished Grade

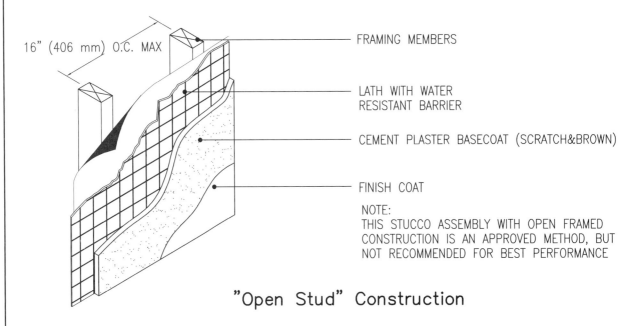

Figure E-1
"Open Stud" Construction

*Source: Northwest Wall & Ceiling Bureau.
Reproduced by permission.*

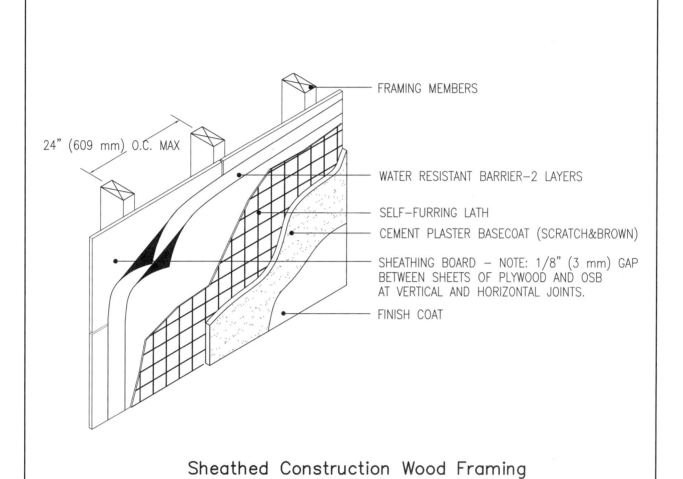

Figure E-2
Sheathed Wood Framing Construction

*Source: Northwest Wall & Ceiling Bureau.
Reproduced by permission.*

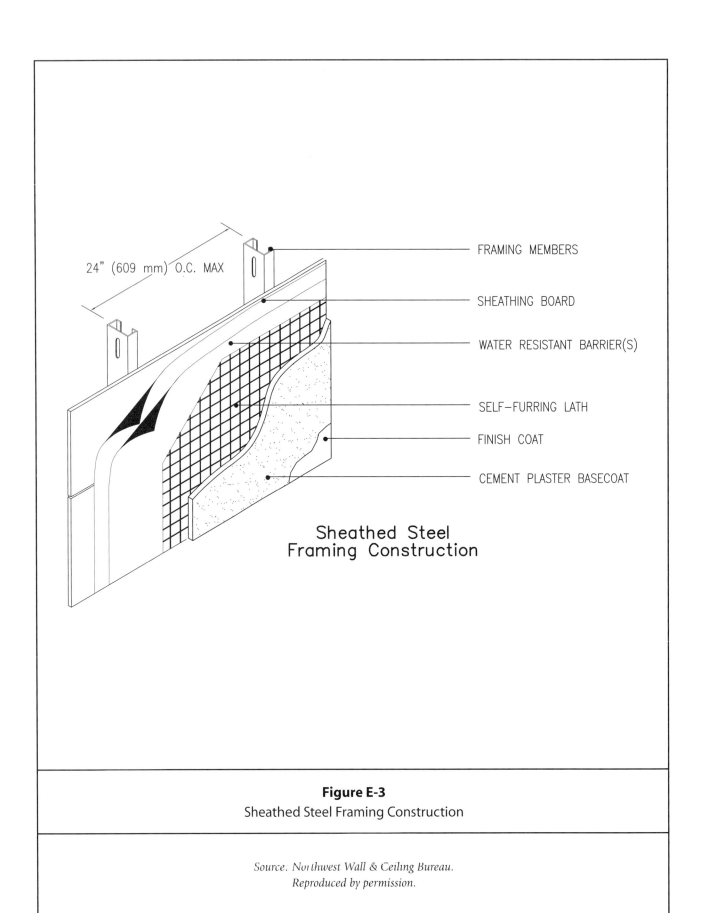

Figure E-3
Sheathed Steel Framing Construction

*Source: Northwest Wall & Ceiling Bureau.
Reproduced by permission.*

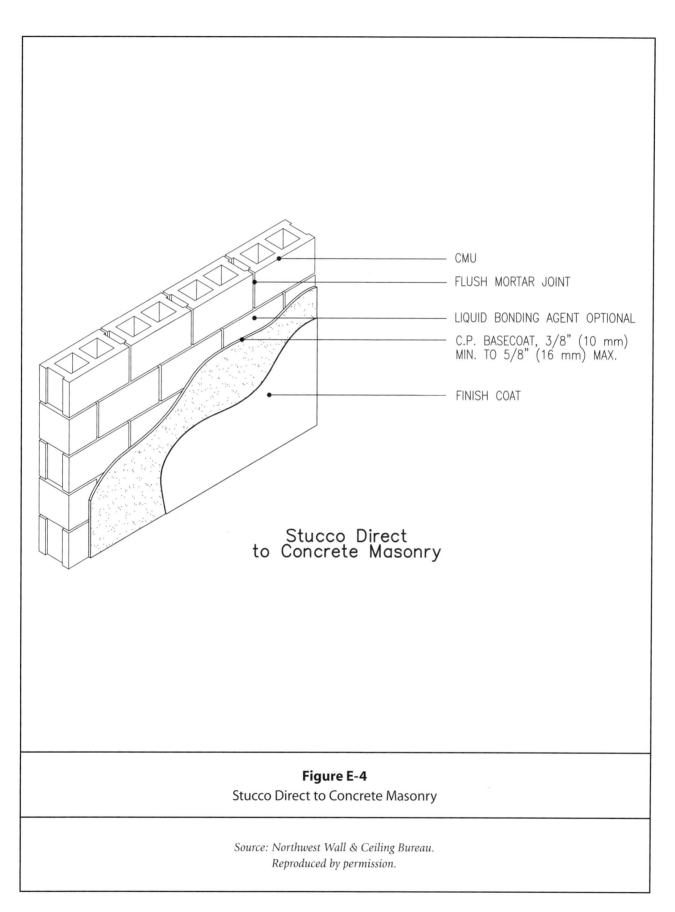

Figure E-4
Stucco Direct to Concrete Masonry

*Source: Northwest Wall & Ceiling Bureau.
Reproduced by permission.*

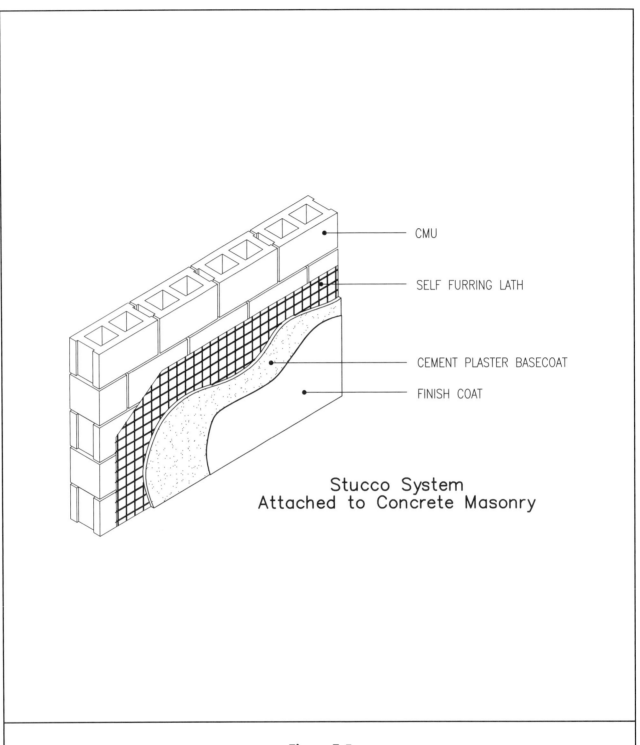

Figure E-5
Stucco System Attached to Concrete Masonry

*Source: Northwest Wall & Ceiling Bureau.
Reproduced by permission.*

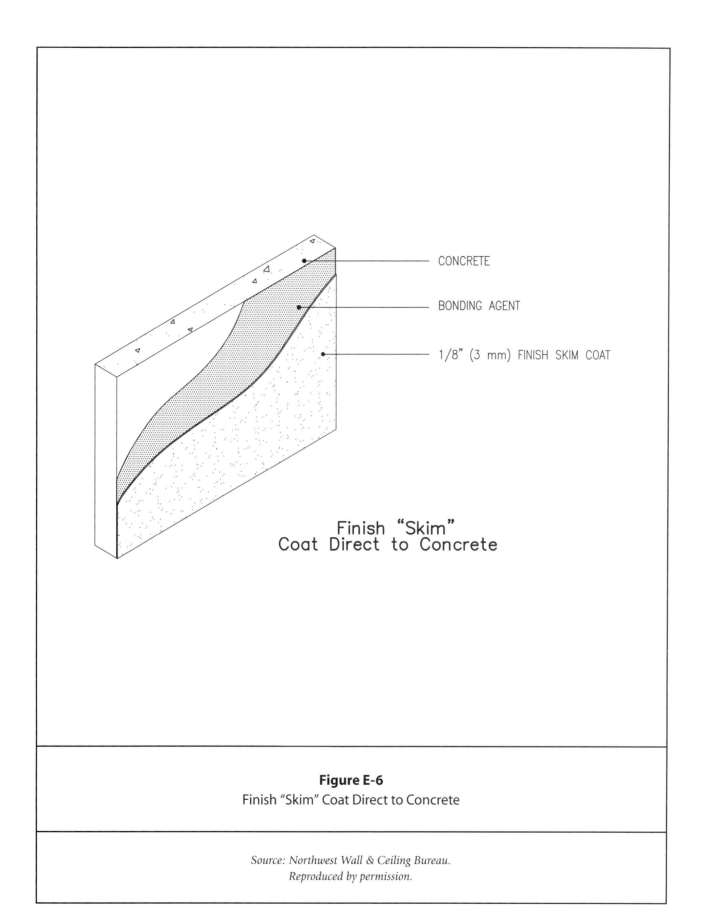

Figure E-6
Finish "Skim" Coat Direct to Concrete

*Source: Northwest Wall & Ceiling Bureau.
Reproduced by permission.*

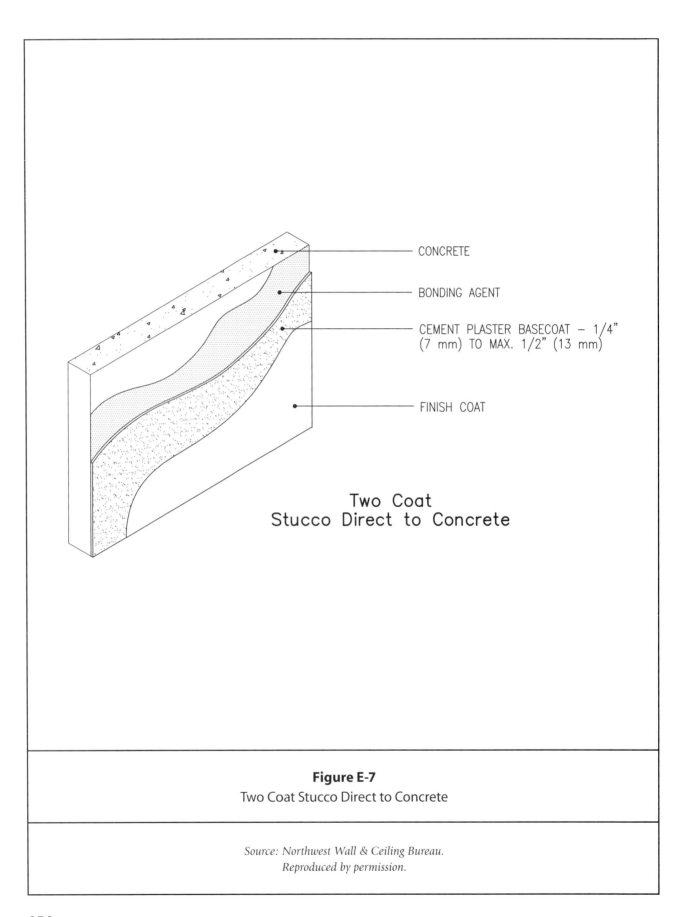

Figure E-7
Two Coat Stucco Direct to Concrete

Source: Northwest Wall & Ceiling Bureau.
Reproduced by permission.

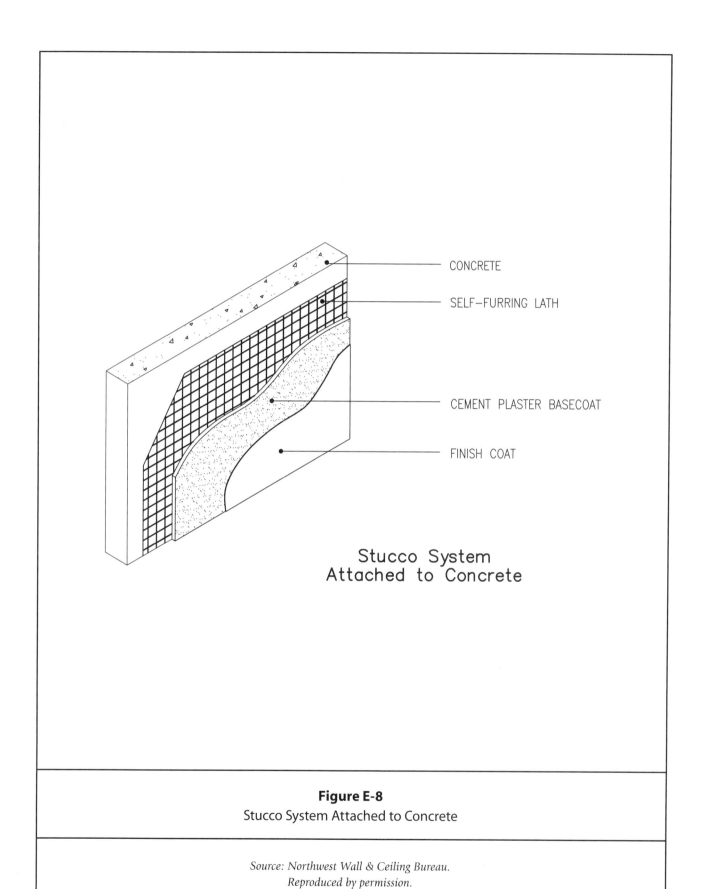

Figure E-8
Stucco System Attached to Concrete

Source: Northwest Wall & Ceiling Bureau.
Reproduced by permission.

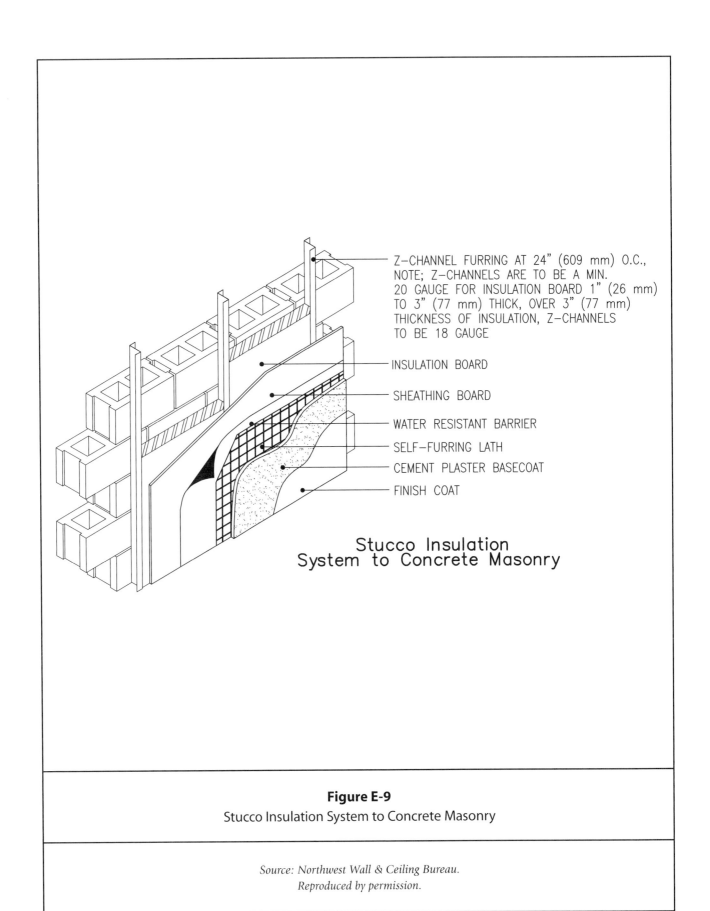

Figure E-9
Stucco Insulation System to Concrete Masonry

Source: Northwest Wall & Ceiling Bureau.
Reproduced by permission.

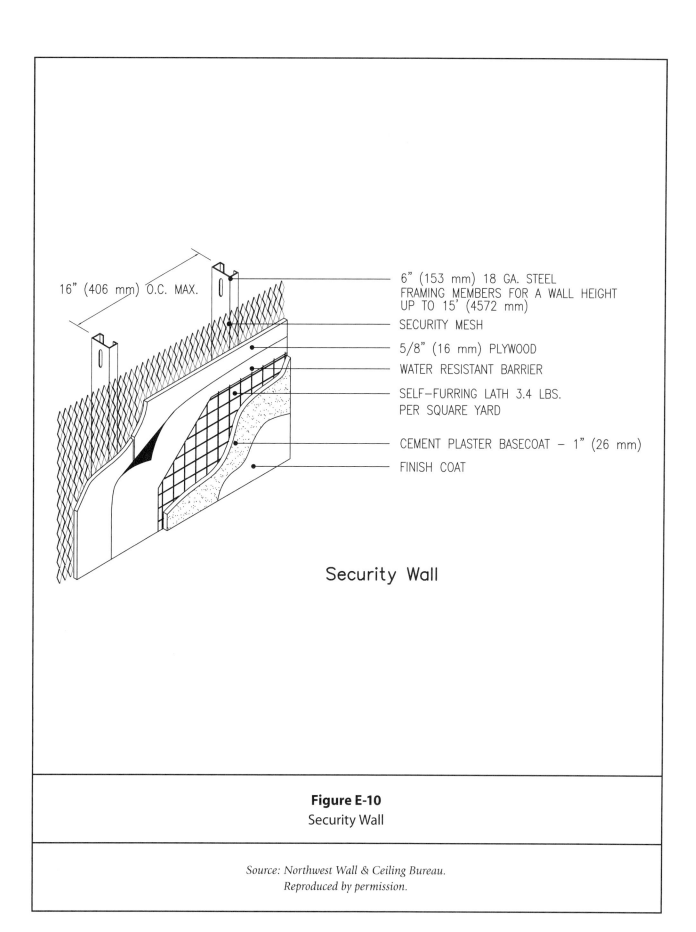

Figure E-10
Security Wall

*Source: Northwest Wall & Ceiling Bureau.
Reproduced by permission.*

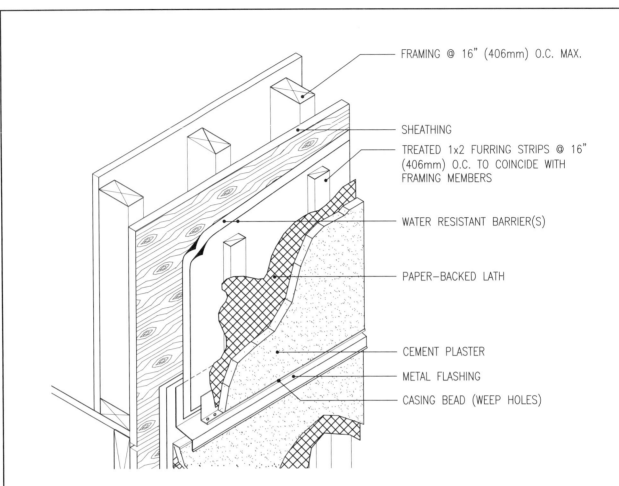

Opened Framed Rainscreen Construction

NOTE 1: The traditional stucco assembly is a concealed weather-barrier system which accomodates water intrusion by means of drainage at the flashing outlets. The rainscreen stucco assembly is an optional method of design. It incorporates a cavity that proivdes an open drainage path for water to get out at the flashing outlets. This mothod also develops the concept of pressure equalization. Proper attention to details, materials, and construction has to be given to a rainscreen design as is given to the stucco concealed weather barrier sytem . The details in this guide manual may not be adaptable to the rainscreen concept.

NOTE 2: This type of stucco system, "open framing", is an approved method, but not recommended for the best stucco wall performance. There is a tendancy for the cement plaster to develop cracks.

Figure E-11
Open Framed Rainscreen Construction

Source: Northwest Wall & Ceiling Bureau.
Reproduced by permission.

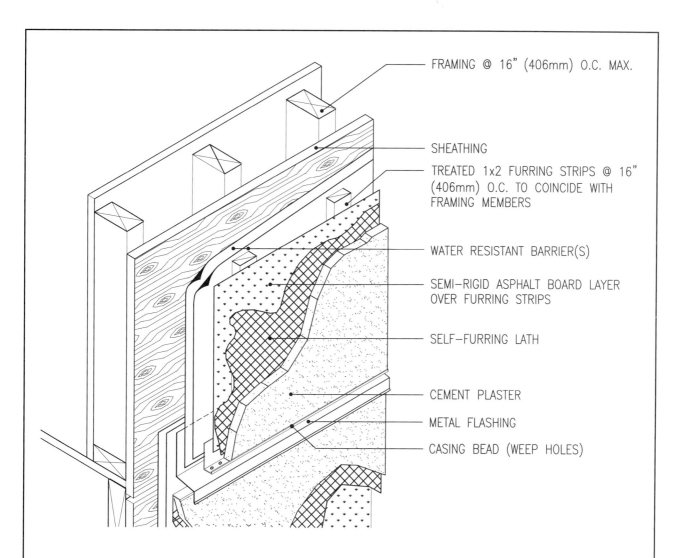

Semi-Rigid Rainscreen Construction

NOTE 1: The traditional stucco assembly is a concealed weather-barrier system which accomodates water intrusion by means of drainage at the flashing outlets. The rainscreen stucco assembly is an optional method of design. It incorporates a cavity that proivdes an open drainage path for water to get out at the flashing outlets. This mothod also develops the concept of pressure equalization. Proper attention to details, materials, and construction has to be given to a rainscreen design as is given to the stucco concealed weather barrier sytem . The details in this guide manual may not be adaptable to the rainscreen concept.

Figure E-12
Semi-Rigid Rainscreen Construction

Source: Northwest Wall & Ceiling Bureau.
Reproduced by permission.

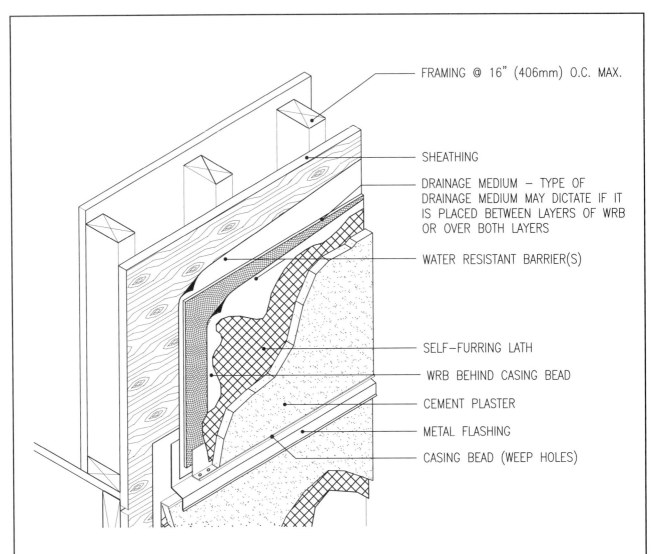

Drainage Medium Construction

NOTE 1: The traditional stucco assembly is a concealed weather-barrier system which accomodates water intrusion by means of drainage at the flashing outlets. The rainscreen stucco assembly is an optional method of design. It incorporates a cavity that proivdes an open drainage path for water to get out at the flashing outlets. This mothod also develops the concept of pressure equalization. Proper attention to details, materials, and construction has to be given to a rainscreen design as is given to the stucco concealed weather barrier sytem . The details in this guide manual may not be adaptable to the rainscreen concept.

Figure E-13
Drainage Medium Construction

Source: Northwest Wall & Ceiling Bureau.
Reproduced by permission.

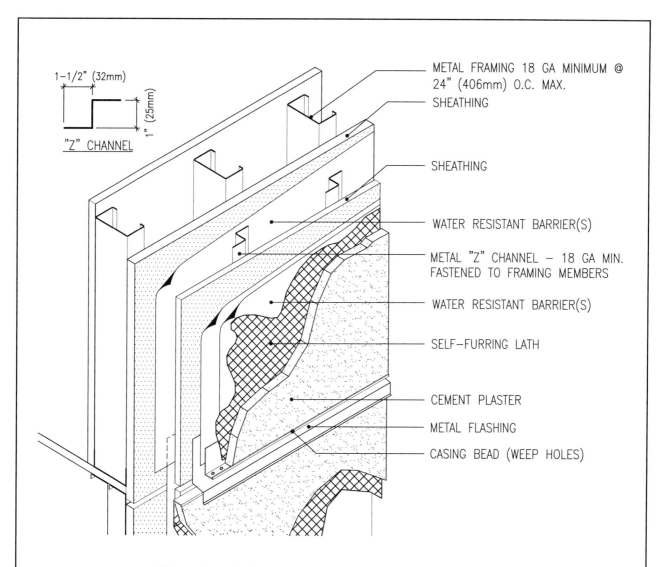

Sheathed Rainscreen Construction

NOTE 1: The traditional stucco assembly is a concealed weather-barrier system which accomodates water intrusion by means of drainage at the flashing outlets. The rainscreen stucco assembly is an optional method of design. It incorporates a cavity that proivdes an open drainage path for water to get out at the flashing outlets. This mothod also develops the concept of pressure equalization. Proper attention to details, materials, and construction has to be given to a rainscreen design as is given to the stucco concealed weather barrier sytem . The details in this guide manual may not be adaptable to the rainscreen concept.

Figure E-14
Sheathed Rainscreen Construction

Source: Northwest Wall & Ceiling Bureau.
Reproduced by permission.

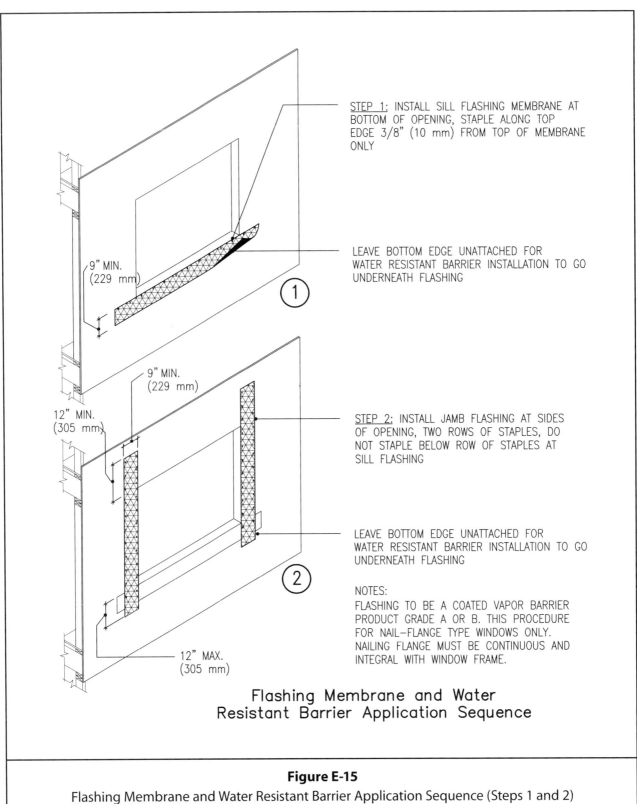

Figure E-15
Flashing Membrane and Water Resistant Barrier Application Sequence (Steps 1 and 2)

Source: Northwest Wall & Ceiling Bureau.
Reproduced by permission.

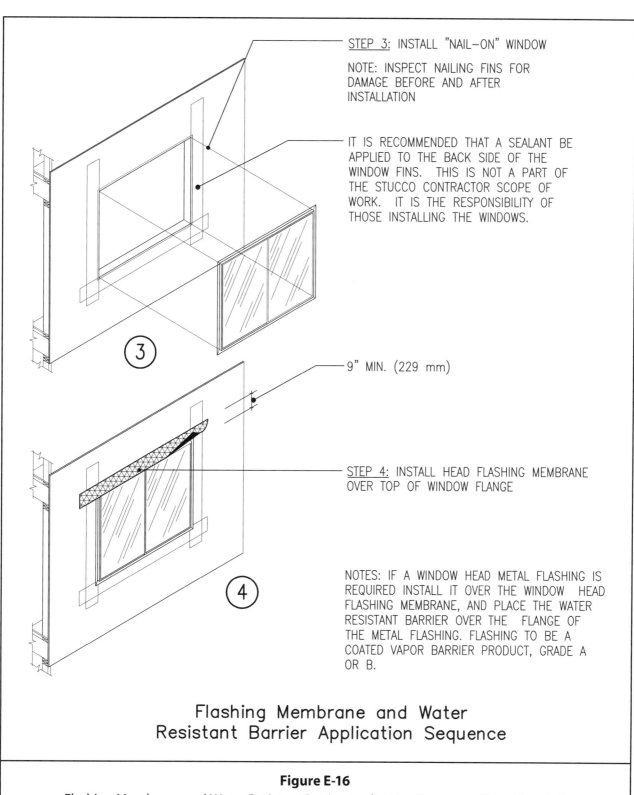

Figure E-16
Flashing Membrane and Water Resistant Barrier Application Sequence (Steps 3 and 4)

Source: Northwest Wall & Ceiling Bureau.
Reproduced by permission.

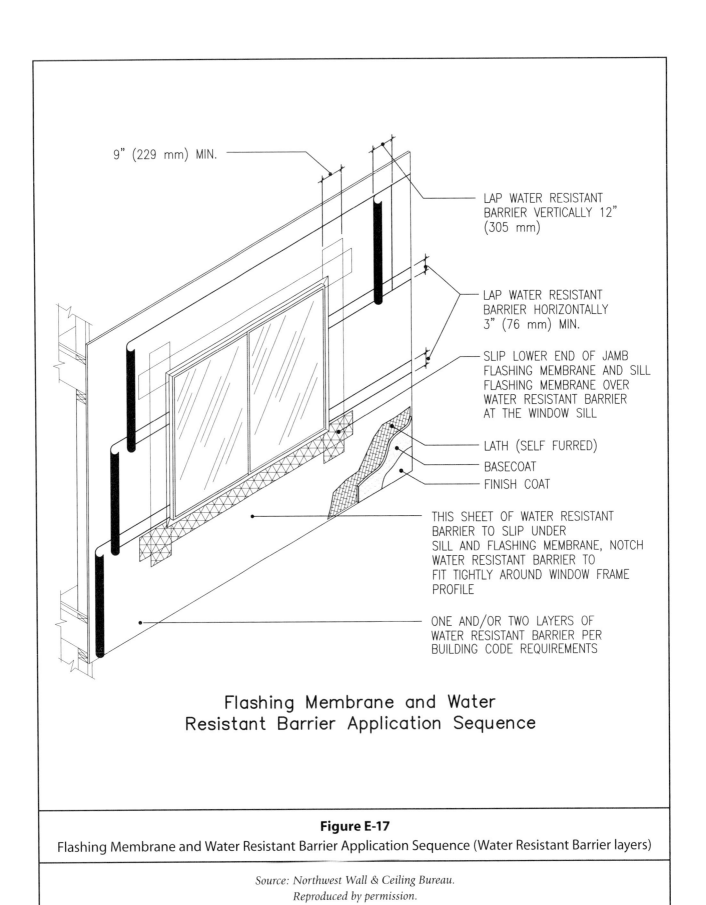

Figure E-17
Flashing Membrane and Water Resistant Barrier Application Sequence (Water Resistant Barrier layers)

Source: Northwest Wall & Ceiling Bureau.
Reproduced by permission.

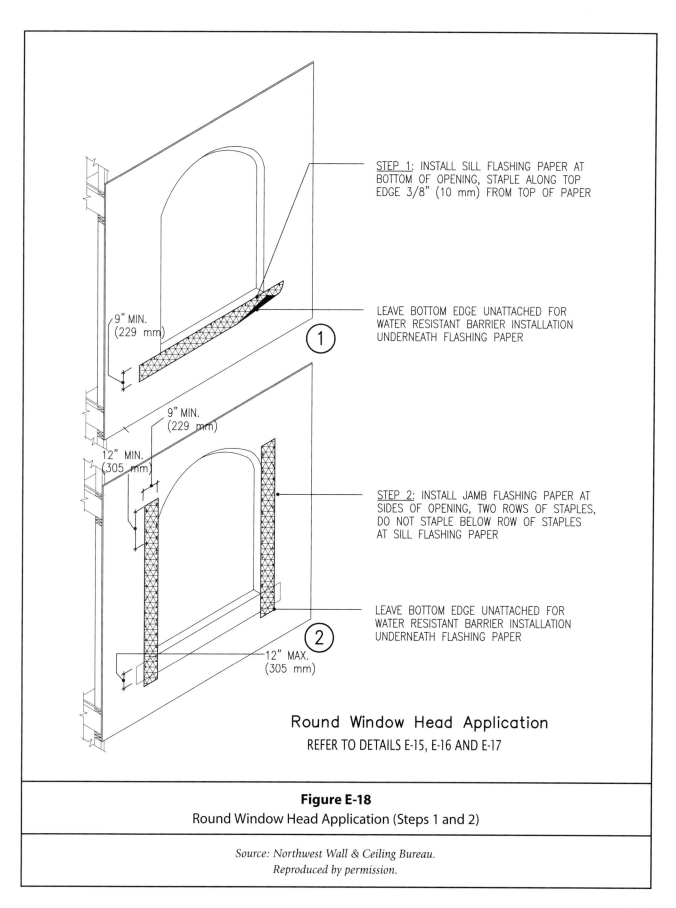

Figure E-18
Round Window Head Application (Steps 1 and 2)

Source: Northwest Wall & Ceiling Bureau.
Reproduced by permission.

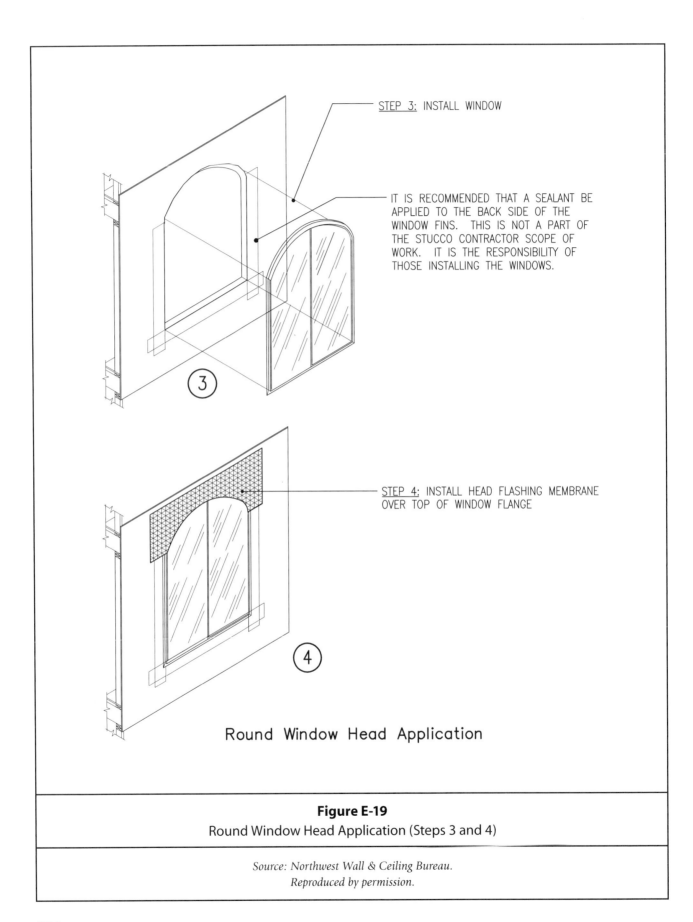

Figure E-19
Round Window Head Application (Steps 3 and 4)

Source: Northwest Wall & Ceiling Bureau.
Reproduced by permission.

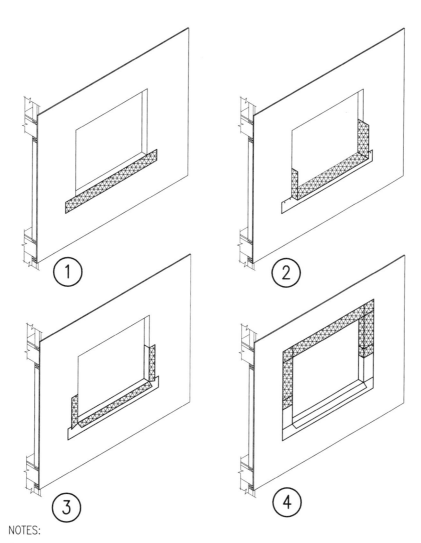

NOTES:

FLASH THE WINDOW OPENING BY INSTALLING MOISTURE BARRIER MEMBRANE AS SHOWN ABOVE BY WRAPPING THE ROUGH INSIDE SURFACE OPENING.

A NAIL ON WINDOW WOULD BE INSTALLED OVER THE FLASHING SYSTEM, AND THEN THE STUCCO WATER RESISTANT BARRIER WOULD BE INSTALLED PER DETAIL E-17

OPTION: INSTALL A SILL PAN WITH UPTURNS.

Flashing Wrapping Rough Inside Opening Application
ALTERNATE METHOD FOR FLASHING ROUGH OPENING AREAS

Figure E-20
Flashing Wrapping Rough Inside Opening Application

Source: Northwest Wall & Ceiling Bureau.
Reproduced by permission.

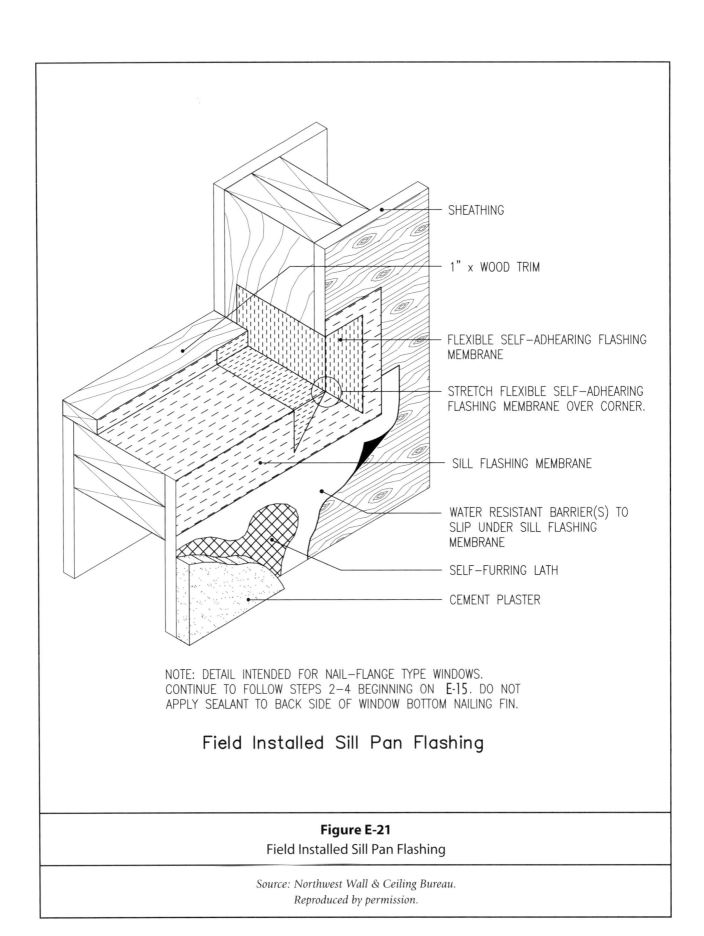

Figure E-21
Field Installed Sill Pan Flashing

*Source: Northwest Wall & Ceiling Bureau.
Reproduced by permission.*

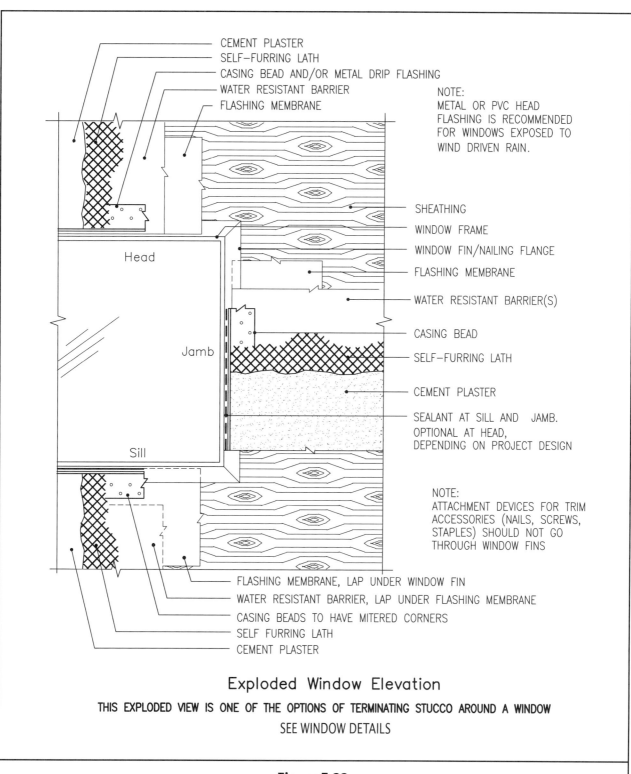

Figure E-22
Exploded Window Elevation

*Source: Northwest Wall & Ceiling Bureau.
Reproduced by permission.*

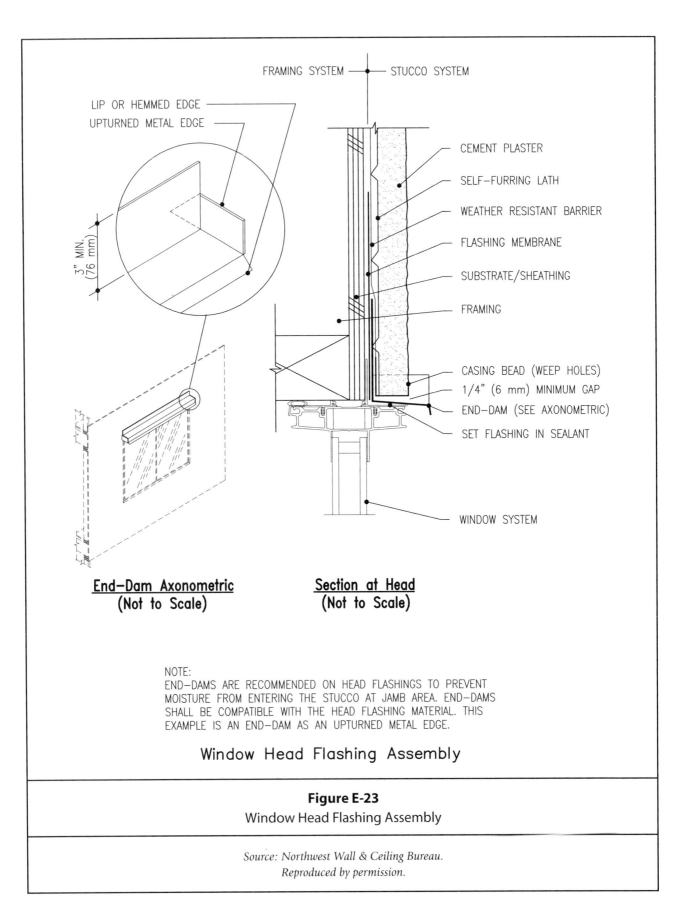

Figure E-23
Window Head Flashing Assembly

Source: Northwest Wall & Ceiling Bureau.
Reproduced by permission.

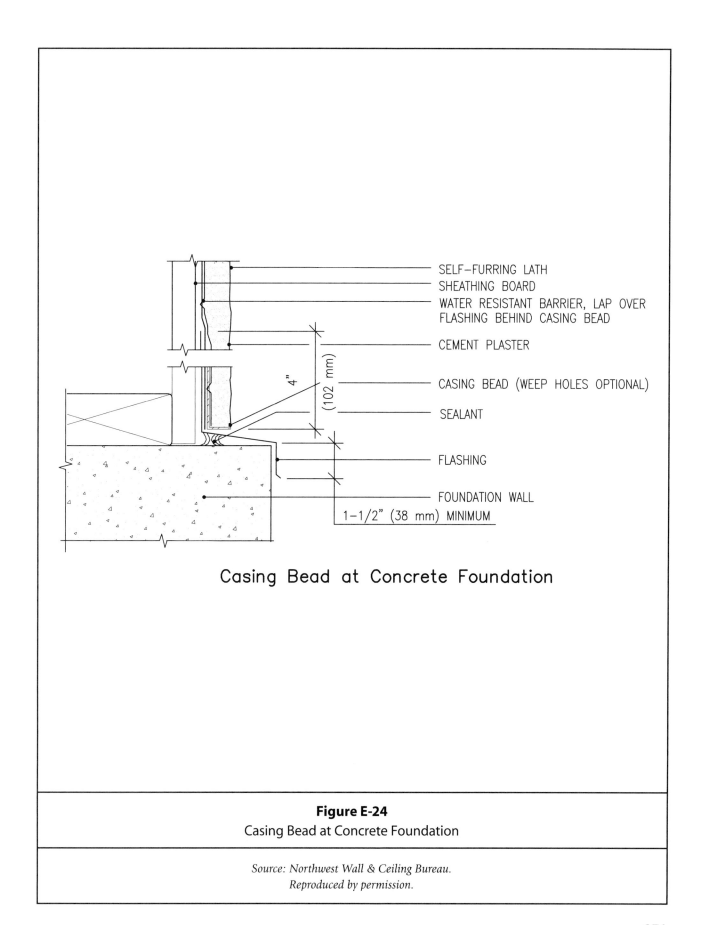

Figure E-24
Casing Bead at Concrete Foundation

*Source: Northwest Wall & Ceiling Bureau.
Reproduced by permission.*

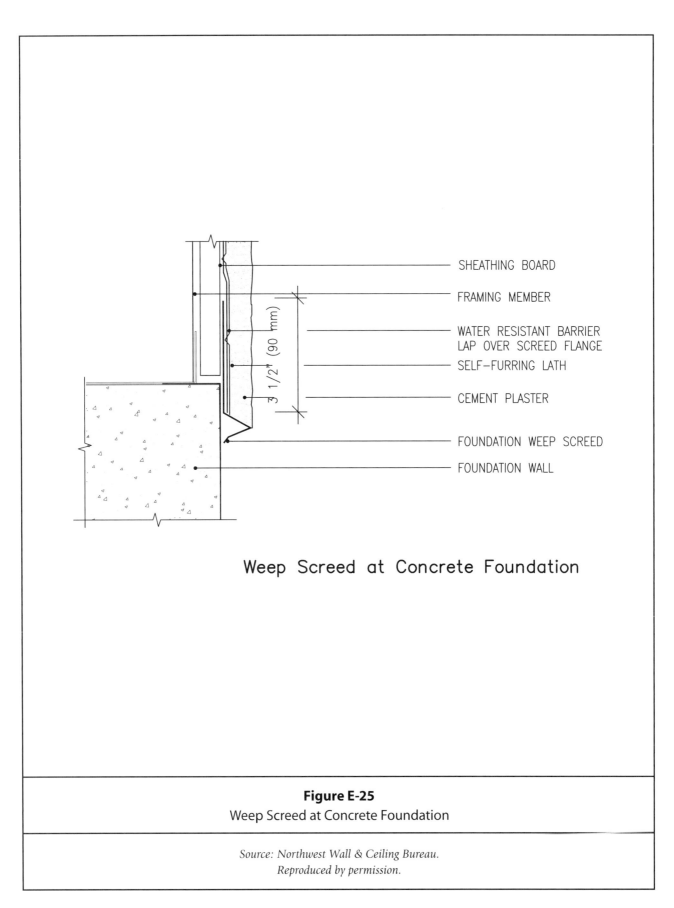

Figure E-25
Weep Screed at Concrete Foundation

*Source: Northwest Wall & Ceiling Bureau.
Reproduced by permission.*

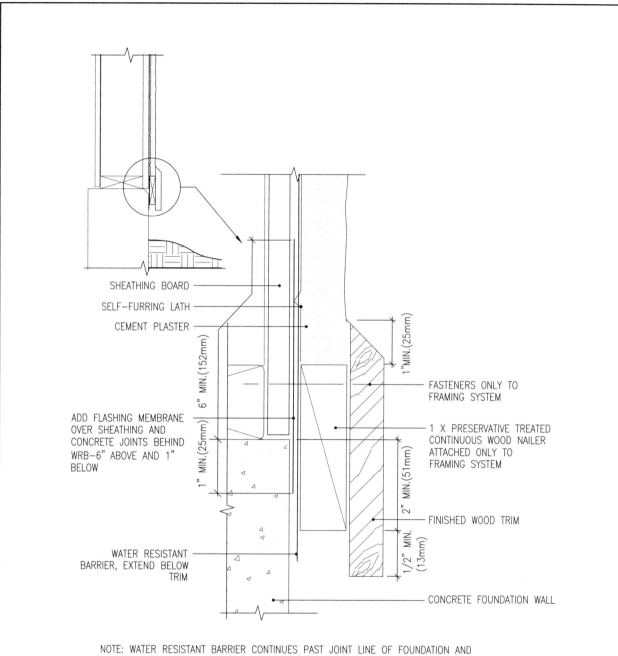

Wood Trim at Concrete Foundation

Figure E-26
Wood Trim at Concrete Foundation

*Source: Northwest Wall & Ceiling Bureau.
Reproduced by permission.*

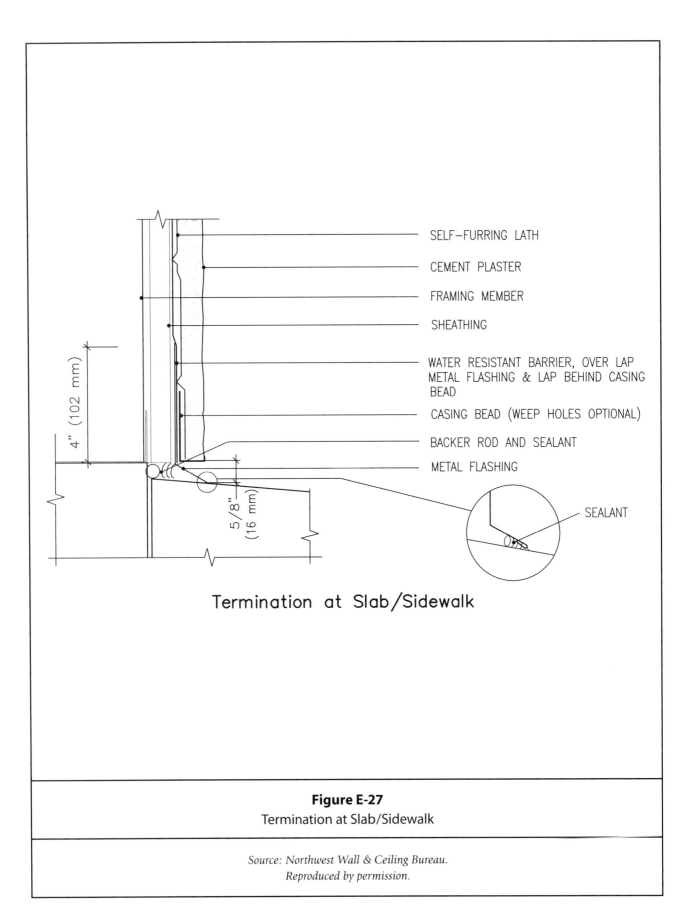

Figure E-27
Termination at Slab/Sidewalk

*Source: Northwest Wall & Ceiling Bureau.
Reproduced by permission.*

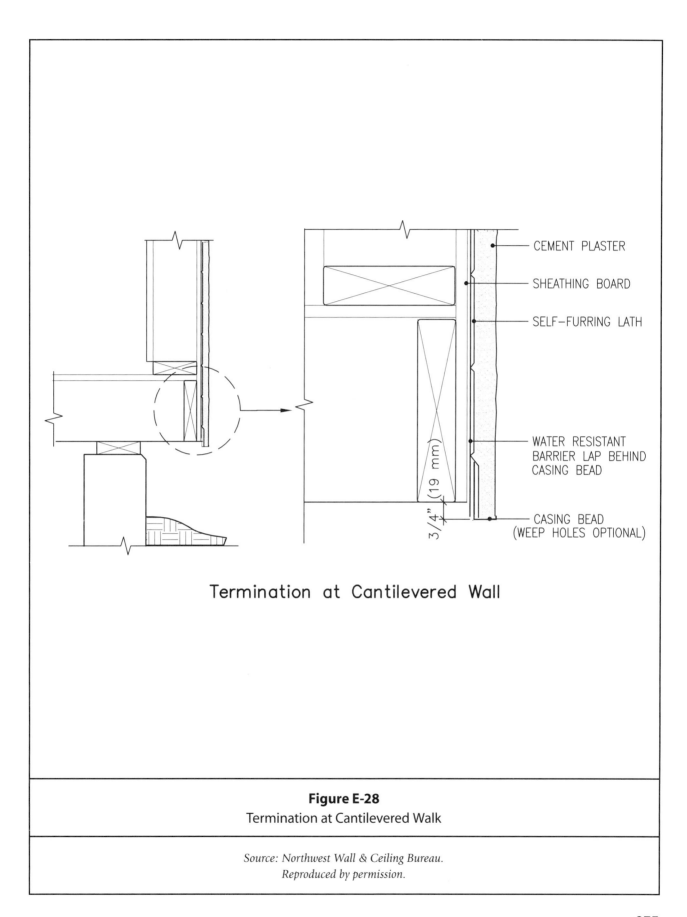

Figure E-28
Termination at Cantilevered Walk

*Source: Northwest Wall & Ceiling Bureau.
Reproduced by permission.*

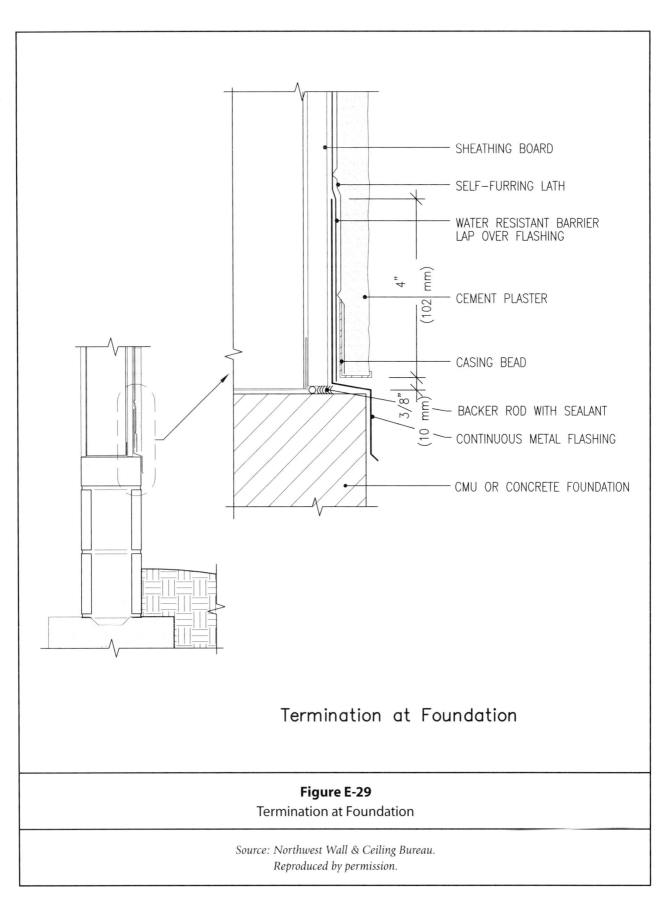

Figure E-29
Termination at Foundation

*Source: Northwest Wall & Ceiling Bureau.
Reproduced by permission.*

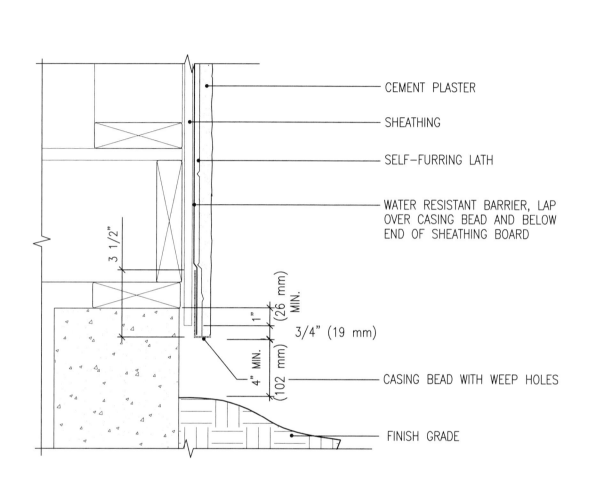

Figure E-30
Termination at Foundation/Finished Grade

Source: Northwest Wall & Ceiling Bureau.
Reproduced by permission.

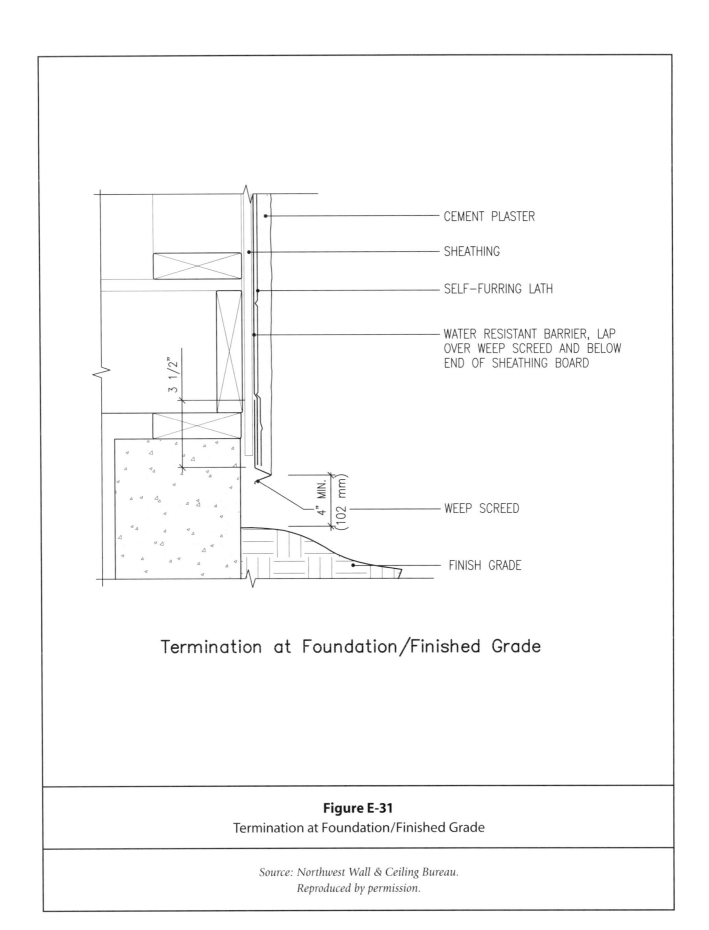

Figure E-31
Termination at Foundation/Finished Grade

*Source: Northwest Wall & Ceiling Bureau.
Reproduced by permission.*

Index

Symbols

#66 beads—28

A

accessories—3, 28
  illustrated—34
  intersections, joining—24
  materials—28, 31

acoustic ceiling tiles—123

acoustic plaster—123

acoustic properties
  plaster properties—122

acoustics
  for typical walls—123

air line—77

application
  brown coat—83
  finish coat—86, 90
  hand—74
  machine—74
  patterns—80, 90
  required tools—71
  scratch coat—81

application patterns—80

architectural effects—102

asphalt paper—43

ASTM C 91—63, 64

ASTM C 144—64

ASTMC 150—63, 64

ASTM C 206—64

ASTM C 332—64

ASTM A 653—28

ASTM C 897—64

ASTM C 926—86

ASTM C 1063—28, 47

ASTM C 1328—63

ASTM D 226 Felt
  alternatives—44

B

Barrier EIFS Systems—57

base—2

base coat—4

bond coat—12, 16

bonding agent—11, 12, 13, 15

bridging
  dissimilar base materials—14

brocade texture—94

brown coat—4
  application—83
  cure—86
  floating (rodding)—85
  thickness—71

building felt—3, 43

building paper—3

C

California—9, 18
  self-furring lath—53

California Building Code—106

casing bead—3, 28, 36
  abutting—41
  at floor line—38
  making expansion joints—30

caulking and sealants—3, 23, 24, 31
  application—32

ceilings—33

changing color—142

channel screeds—30

checklist
  preparation—73
  Stucco Application Checklist (residential)—6

CMU—15

coats
  thickness—71

coefficients of heat transmission—120

color
  built-in—98
  changing—142

color coat—5

coloring compounds—98
  finish coat—89

concrete—11
  moderately rough—12
  problem—13
  smooth—11, 12

concrete building—2

concrete masonry building—2

concrete masonry unit (CMU)—15

control joints—14, 29, 32
  abutting at corners—37
  coordinating with lath—53
  in concrete or masonry walls—11
  intersections—37
    with weep screed—38
  wall areas—33

corners—3
  inside—41

corner aid
  outside corners—41

corner bead—29, 36
  outside—41

corner lath—30

curing—5, 14, 16

D

damp-proofing—121

darby—85

dash coat—12, 16

decorative stucco—101

decorative tile inset—104

design considerations—5

diamond mesh lath—51, 52

dimple lath—52

double-V joint—29

ducts—41

E

earthquake—9, 105, 106, 109

EIFS—3, 55, 102
- advantages—55
- Barrier EIFS Systems—57
- history—55
- installation—58
- maintenance and repair—59
- materials—55, 56
- One-Coat Base Coat System—3
- system approach—56
- Wall Drainage EIFS Systems—57

EIFS installation—58

elastomeric acrylic coatings—99

electric meters—42

end dams—23, 24

English texture—91

EPS Corner implant—104

EPS foam shapes—102

esitmating
- shores—129

estimating—125
- craft direct cost—131
- general contractor's direct costs—132
- hourly labor costs—129
- indirect job costs—133
- lathing—130
- regional modifiers—133
- rentals—134
- Steel Scaffolding and Accessories—127

expanded metal lath—51

expansion joint, weeping
- at floor line—39

expansion joints—30
- abutting at corners—37
- intersections—37

Exterior Insulation and Finish Systems (EIFS)—3

F

F-reveals—30

fascia 40

fasteners
- powder-actuated—50
- power—50

finish coat—5, 89, 143
- application—86
- application pattern—90
- coloring compounds—89
- mixes—68, 69
- mixing—67, 68
- sand—89
- thickness—72, 89

fire-resistance
- plaster properties—113

fireproofing columns—117

flashing—22, 46
- intersections, joining—24
- kickout—26
- window—25
  - head—26

floating—85

Fortifiber Jumbo Tex—44

frame building
- wood—20
- wood or metal—18

furring—3
- z-channel—17

G

Grade D Kraft building paper—44
- alternatives—45

ground—36

H

hand application—74

head flashings
- windows—22

housewrap—3, 44

humidity—4

hydration—65

I

implant—101, 102
- decorative tile inset—104
- EPS Corner—104

installation
- EIFS—58

insulation
- plaster properties—112

interior stucco—1

J

J channel—28

Jumbo Tex—44

K

key—4

kickout flashing—26
- coordination with roofing—27

knockdown dash texture—93

L

lateral (horizontal) forces—9, 105, 110

lateral reinforcement—3, 109, 110

lath—3
- coordinating with control joints—53
- diamond mesh—51, 52
- dimple—52
- expanded metal—51
- paper-backed—47
- paper-backed woven wire fabric—51
- plain wire fabric—52
- ribbed—51
- riblath—51
- self-furring—51, 53
- self-furring diamond mesh (dimple)—52
- self-furring diamond mesh (groove)—52
- separate installation—48
- strip—52
- welded wire—51
- wood—54
- woven wire fabric—50

lath attachment—48
- on concrete—14
- to concrete and masonry—50
- to metal framing—50
- to wood framing—49

light dash texture—94

lime—64

line wires—18
- tightening—18

local codes and ordinances—5, 43, 45, 106
- one-coat stucco systems—60

Los Angeles Building Code—106

M

machine
  clean-up and maintenance—79

machine application—74

machine placement—75

machine setup—75
  priming pump—78
  pump and nozzle—76

maintenance and repair—135
  cleaning stucco—140
  EIFS—59

manufacturers' documentation—5

Marblecrete texture—96

masonry—11, 15
  mortar joints—15

masonry surfaces
  types—15

material choices—5

medium dash texture—93

metal frame buildings—2
  rainscreen construction—21

metal lath
  self-furred—13

mix
  general proportions—1, 65

mixes
  bagged—66
  base coats—68
  finish coat—68, 69
  Portland cement plaster—63

mixing
  finish coat—67, 68
  guidelines—67
  hand—65
  machine—65
  steps—66

Monterey texture—92

mortar joints—15
  brittle—15
  struck-flush—15

multi-story buildings—9, 18

multiple coats—1, 4

N

No. 15 felt—44

No. 40 joint—30
  at floor line—38

Northridge Earthquake—9, 105, 109

nozzle—77

O

One-Coat Base Coat System—3

one-coat stucco systems—60
  local codes and ordinances—60
  performance problems—60

open frame construction—2, 18

OSB—2

P

painted concrete—11

painting stucco—142
  cement-based paints—98
  fog coat—98, 143
  latex paint—98
  reasons—98

paper-backed lath—47
  shingle pattern—47

paper-backed woven wire fabric lath—51

parapets—40
  intersecting with walls—40

permeability
  weather-resistant barrier—44

pipes—41

plain wire fabric lath—52

plant-on—101, 102

plaster properties
  acoustic properties—122
  fire-resistance—113
  insulation—112
  structural strength—105
  thermal properties—112
  water-resistance—121

plasticizer—64

plywood—2, 9

porosity—43

Portland cement
  Type I—63
  Type II—64
  Type III—64
  Type IV—64
  Type V—64
  white—63, 89
  mixes
    bagged—66
    base coats—68
    finish coat—68, 69
    Portland cement plaster—63

Portland cement plaster
  mixes—63
  mixing
    finish coat—67, 68
    guidelines—67
    hand—65
    machine—65
    steps—66

power-actuated fasteners—14

R

rain
  wind-driven—45

rainscreen construction—20, 21

reentrant corners—30

Renaissance texture—95

repairs—135
  changing colors—142
  cracks—139
  EIFS—59
  paper and lath—137
  patching stucco—137
  wood lath—54

required tools
  application—71

restoration
  wood lath—54

reveals—30

ribbed lath (rib lath)—51

rodding—85

roofing
  kickout flashing—27

roof line—40

rustication—30

S

safety precautions—65

sand—64
   finish coat—89

Saturated/Surface Dry (SSD)—11, 12

scaffolding—10

Scope of Work Responsibilities—126

scratch coat—4
   application—81
   cure—83
   thickness—71

scratching—4

screed—36

sealed concrete—11

self-furring—3

self-furring diamond mesh lath (groove)—52

self-furring fastener—19

self-furring lath—3, 19, 51, 52. 53

shear strength—9, 105, 110

sheathed construction—19, 110

sheathing—9, 18
   OSB—2, 18
   plywood—2, 18
   choice of materials—19

shingle pattern—3, 22, 46
   paper-backed lath—47

silicone coatings—99

sill pans—24

skim coat—11

soffits—40

soffit vents—30, 40

Spanish texture—92

sprayed acoustic plaster—123

SSD—11

stop bead—28

strip lath—52

struck-flush mortar joints—15

structural strength
   plaster properties—105

stucco
   defined—1
   exterior—1
   interior—1
   porosity—43

Stucco Application Checklist (residential)—6

stucco damage
   identifying—135

stucco failure
   causes—136

StuccoWrap—45

subsurface—2

summer heat—5

surface preparation—2

T

temperatures (ambient)—4, 72
   summer heat—5, 72

texture coat—5

texture finish choices—90

textures
   brocade—94
   broomed—97
   brushed—97
   English—91
   imprinted—97
   knockdown dash—93
   light dash—94
   Marblecrete—96
   medium dash—93
   Monterey—92
   Renaissance—95
   scratched—97
   Spanish—92
   stamped—97
   stippled—97
   techniques for achieving—96
   travertine—91, 101
   trowel sweep—95

thermal properties
   plaster properties—112

thickness
   brown coat—71
   coats—71
   finish coat—72
   scratch coat—71

three-coat system—1

tie wire—3, 31
   material—31
   tying configurations—31

timing—4

travertine texture—91, 101

trowel sweep texture—95

Typar—44, 45

Type I
   Portland cement—63

Type II
   Portland cement—64

Type III
   Portland cement—64

Type IV
   Portland cement—64

Type V
   Portland cement—64

Tyvek—44, 45

V

vent covers—42

vents—41

W

Wall Drainage EIFS Systems—57

water—64

water-resistance
   plaster properties—121

water-resistant barrier—3
   See weather-resistant barrier

water-resistive barrier—3
   See weather-resistant barrier

waterproofing—121

weather-resistant barrier—3, 43, 121
   double layer—45
   permeability—44
   single layer—45

weather conditions—4, 72

Weathermate Plus—44, 45

weep screed—3, 28, 36, 40, 42

welded wire lath—51

whip line—77

white Portland cement—63, 89

wind—9, 106

wind-driven rain—45, 121

window flashing—25
    head—26

windows
    casing bead—39
    control joint at head corner—39
    control joint at window sill corner—39
    head corners—39
    sill corners—39

winds—4

wood frame buildings—2
    sheathed—20
        rainscreeen construction—20

wood lath—54

woven wire fabric lath—50

Z

z-channel—21
    furring—17

BUILDER'S GUIDE TO
STUCCO
Lath & Plaster

Using the PDF Version

BASIC USAGE

To use the PDF files on the enclosed CD-ROM, you need the Adobe Reader (available for free download from http://www.adobe.com/products/acrobat/readstep2.html). As the size of the PDF file for the book is quite large, you may achieve better performance (and find it more convenient) to copy the PDF file to your hard drive and access it from there. (Under the Limited License below, you are allowed to copy this PDF file to a maximum of two (2) personal computer hard drives which you use for your personal or business research.)

Once you open the PDF file, you can navigate the book by using the links in the left-hand list of "bookmarks." There are also links in the book's Table of Contents and Index. You can also use Adobe Reader's Search capabilities, to find pages containing words or phrases that interest you.

IMPORTANT NOTES ABOUT PRINTING

- As the CD-ROM is provided as a companion to the printed book, **printing from the PDF Version has been disabled for the main book**.

LIMITED LICENSE / TERMS AND CONDITIONS OF USE

This PDF Version is provided as an electronic version of the printed edition of Builder's Guide to Stucco, Lath & Plaster (Book), under a limited, non-exclusive license: You (the original purchaser) may use the PDF Version to conduct research for personal or business purposes.

You may install/copy the PDF file to a maximum of two (2) personal computer hard drives, which personal computers you use for your personal or business research. Copying and printing capabilities have been restricted/disabled on the PDF version of the Book; nonetheless, you may not print, copy, reproduce or otherwise duplicate or distribute any portion of this PDF Version (including but not limited to the entire PDF file or any pages or text therefrom) for others' use. You are specifically prohibited from loaning, sharing, copying or otherwise transferring this CD-ROM disc and/or any of the files contained therein to third parties. If you do not agree with all of these terms and conditions, you are hereby prohibited from using the PDF Version.

NOTICES TO THE READER

The Publisher has made every effort to provide complete and accurate information, but does not guarantee the accuracy or completeness of any information published herein, nor shall the Publisher have neither liability nor responsibility to any person or entity for any errors, omissions, or damages arising out of use of this information. Builder's Guide to Stucco, Lath & Plaster is published with the understanding the Publisher is not attempting to render professional services. If such services are required, the assistance of an appropriate professional should be sought. For future updates, errata, amendments and other changes contact Builder's Book, Inc. (the Publisher).

The information contained in this publication is subject to change without notice. All rights reserved. No part of this book may be reproduced or utilized in any form or by any means, electronic or mechanical, including photocopying, recording or by any information storage and retrieval systems, without special permission in writing from the publisher.

Copyright 2007 Builder's Book, Inc. All rights reserved.

Builder's Book, Inc.
BOOKSTORE • PUBLISHER
8001 Canoga Avenue / Canoga Park, CA 91304
1-800-273-7375 / www.buildersbook.com